Practical Oscillator Handbook

Irving M. Gottlieb PE

Newnes
An imprint of Butterworth-Heinemann
Linacre House, Jordan Hill, Oxford OX2 8DP
A division of Reed Educational and Professional Publishing Ltd

A member of the Reed Elsevier plc group

OXFORD BOSTON JOHANNESBURG
MELBOURNE NEW DELHI SINGAPORE

First published 1997

British Library Cataloguing in Publication Data

A catalogue record for this book is available from the British Library

ISBN 0 7506 6312 3

Library of Congress Cataloguing in Publication Data

A catalogue record for this book is available from the Library of Congress

Typeset by Vision Typesetting, Manchester
Printed and bound in Great Britain by
Biddles Ltd, Guildford and King's Lynn

Contents

Preface

The subject of oscillators has been somewhat of a dilemma; on the one hand, we have never lacked for mathematically oriented treatises—the topic appears to be a fertile field for the 'long-haired' approach. These may serve the needs of the narrow specialist, but tend to be foreboding to the working engineer and also to the intelligent electronics practitioner. On the other hand, one also observes the tendency to trivialize oscillator circuits as nothing more than a quick association of logic devices and resonant circuits. Neither of these approaches readily provides the required insights to devise oscillators with optimized performance features, to service systems highly dependent upon oscillator behaviour, or to understand the many trade-offs involved in tailoring practical oscillators to specific demands. Whereas it would be unrealistic to infer that these two approaches do not have their place, it appears obvious that a *third* approach could be useful in bringing theory and hardware together with minimal head-scratching.

This third approach to the topic of oscillators leans heavily on the concept of the universal amplifier. It stems from the fact that most oscillators can be successfully implemented with more than a single type of active device. Although it may not be feasible to directly substitute one active device for another, a little experimentation with the d.c. supply, bias networks, and feedback circuits does indeed enable a wide variety of oscillators to operate in essentially the same manner with npn or pnp transistors, N-channel of P-channel JFETs, MOSFETs, op amps or ICs, or with electron tubes. Accordingly, this book chooses to deal with basic operating principles predicated upon the use of the universal active-device or amplifier. This makes more sense than concentrating on a specific device, for most oscillator circuits owe no dependency to any single type of amplifying device.

Once grasped, the theory of the general oscillator is easily put to practical use in actual oscillators where concern must be given to the specific active device, to hardware and performance specifications, and to component values. To this end, the final section of the book presents numerous solid-state oscillators from which the capable hobbyist and practical engineer can obtain useful guidance for many kinds of projects.

It is felt that the reader will encounter little difficulty acclimatizing to the concept of the universal amplifier, for it is none other than the triangular symbol commonly seen in system block diagrams. Although it hasn't been widely used in conjunction with other circuit symbols, the combination works very well with oscillators. It is respectfully submitted that this book will thereby serve as a unique format for useful information about oscillators.

The symbol used for a.c. generator is usually assumed to be a constant voltage generator, i.e., with zero internal resistance. However, in many instances in this book, it must be assumed to be a constant current generator, or at least to have a high internal resistance. For example, this is the case in Fig. 1.41, where if the generator shown is an ideal voltage generator, it will short out L_1. This will alter circuit operation and make the quoted formula for f_0 wrong. It is recommended therefore that the reader bear this in mind when presented with an a.c. generator in this book.

Irving M. Gottlieb PE

1 Frequency-determining elements of oscillators

A good way to understand oscillators is to view them as made up three essential sections. These are:

- the frequency-determining section
- the active device
- a source of d.c. power

The validity of this viewpoint does not require that the three sections be physically separate entities. This chapter will treat the characteristics of the elements involved in the frequency-determining section.

Parallel-tuned LC circuit

Academically and practically, the parallel LC arrangement known as a 'tank' circuit is the most important element for us to become familiar with. In its simplest and most frequently encountered form, it is made up of a single inductor and a single capacitor. Whether or not we desire it, the inevitable 'uninvited guests', a number of dissipative losses, are always present. (See Fig. 1.1.) In the circuit, these losses behave as resistances. Their presence can, indeed, be closely simulated by simple insertions of resistance into the tank circuit. In Fig. 1.2 we see a possible way in which this can be done. This is the most convenient method and will be used frequently in the equations for computing the various tank circuit quantities.

Losses in a tank circuit

Different losses predominate under different situations. In general, the higher the frequency, the greater the radiation loss. Magnetic hysteresis is

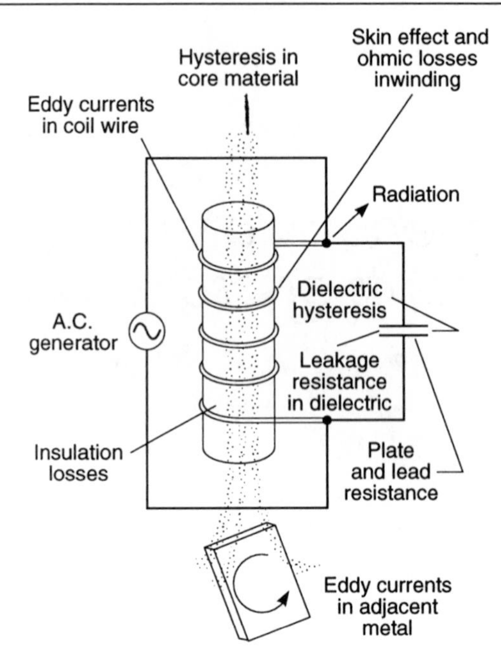

Fig. 1.1 *Some possible losses in an LC tank circuit*

only of consequence when a ferromagnetic core is used, such as powdered iron. The losses due to eddy currents are, in reality, brought about by transformer action in which the offending material constitutes a short-circuited 'secondary'. This being true, we must expect eddy-current losses in the cross-section of the coil winding itself. Skin effect is an a.c. phenomenon that causes the current to concentrate near the surface of the conductor. This is because the more central regions of the conductor are encircled by more magnetic lines than are the regions closer to the surface (see Fig. 1.3). The more lines of magnetic force encircling a conductor, the greater the inductance of the conductor. Hence, the central regions of a conductor carrying alternating current offer higher inductive reactance to the flow of current.

The higher the frequency, the more pronounced is this effect; that is, the greater the tendency of current to concentrate at or near the surface, thereby reducing the effective cross-section of the conductor. Because of skin effect, the resistance offered to the passage of high-frequency current is much higher than the d.c. resistance. (Inductance does not affect the flow or distribution of d.c.) We are not surprised that skin-effect losses are reduced by using hollow conductors of copper content equal to small gauge wire, but which possess a much greater surface area. Also, stranded wire offers more surface for high-frequency conduction than does its 'd.c. equivalent'

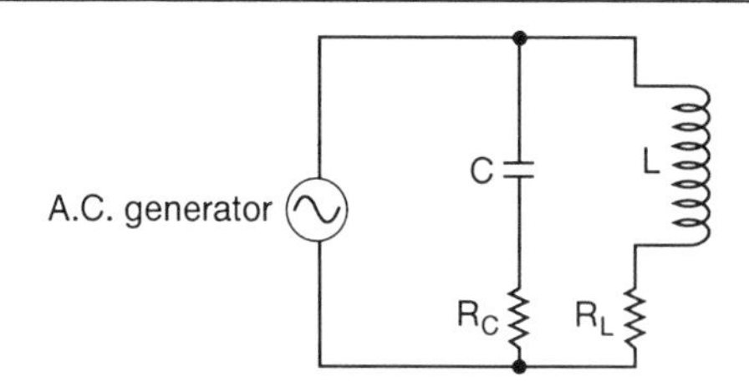

Fig. 1.2 *Representation of losses in an LC tank circuit by series resistances* R_L *and* R_C

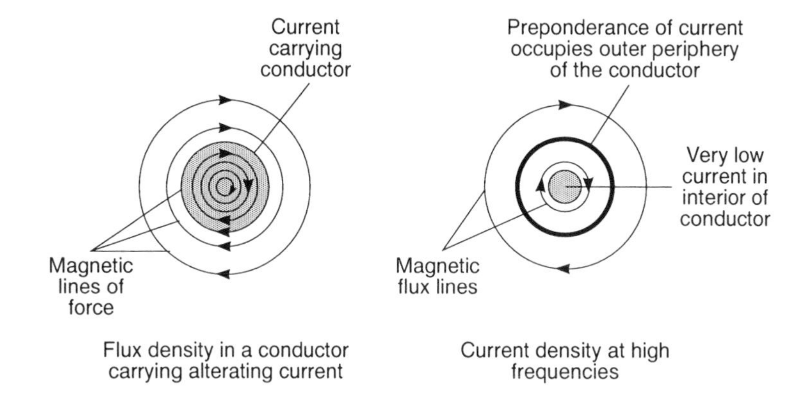

Fig. 1.3 *Flux-density from A.C. in a conductor and the high frequency skin-effect. At low frequencies most of the current flows throughout the cross-section of the conductor. At high frequencies, almost all current is in the outer 'skin' of the conductor*

in solid wire. Stranded wire with each individual wire insulated (Litz wire) is particularly well suited for the flow of high-frequency current.

Dielectric hysteresis in insulating materials is the electrostatic counterpart of magnetic hysteresis in magnetic materials. A frictional effect is displayed by the polarized molecules when they are urged to reverse their charge orientation under the influence of an alternating electric field. There are other losses. Those described and those shown in Fig. 1.1 are, however, the most important. Significantly, in many applications, only the losses in the inductor are of practical consequence, for capacitors often have negligible losses from the standpoint of many practical oscillator circuits.

Characteristics of 'ideal' LC resonant circuit

We find ourselves in a much better position to understand the proprieties of an actual 'lossy' tank circuit by first investigating the interesting characteris-

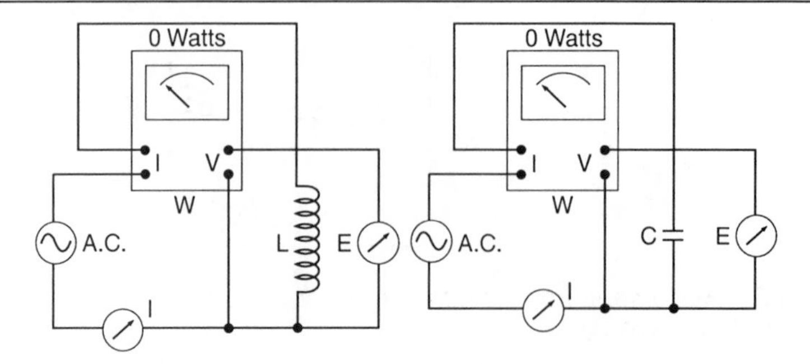

Fig. 1.4 *Voltage and current in an ideal inductor and capacitor*

tics displayed by an 'ideal' tank circuit in which it is postulated that no losses of any kind exist. It is obvious that such an ideal tank circuit must be made up of an inductor and a capacitor that, likewise, have no losses. In Fig. 1.4, we see the important feature of such ideal elements, i.e., when an a.c. voltage is impressed across an ideal inductor or an ideal capacitor, current is consumed, but no power is dissipated. Although there is current through these elements, and voltage exists across them, the wattmeters show a zero reading. This may seem strange at first; such a situation is the consequence of the 90° difference in phase between voltage and current. This phase condition is shown in Fig. 1.5 for the ideal conductor, and in Fig. 1.6 for the ideal capacitor.

In both instances, power is drawn from the source for a quarter cycle, but is returned to the source during the ensuing quarter cycle. This makes the power frequency twice that of the voltage or current waves. This need not be cause for surprise, since the same situation prevails for a resistance energized from an a.c. source. It turns out that the double-frequency power curve is of little practical consequence as such. Of great importance is the fact that the negative portions of the power curves in Figs 1.5 and 1.6 represent power returned to the source; conversely, in the resistance circuit of Fig. 1.7 we note there are no negative portions of the power curve. (All the power drawn by the resistance is dissipated as heat and/or light; *no* power is returned to the source at any time.)

Negative power

We observe in Figs 1.5 and 1.6 that sometimes the voltage is positive when the current is negative and vice versa. By the algebraic law of signs (the product of quantities having unlike signs yields a negative number) it is just such occurrences that produce the negative excursions of the power waveform. Also, every time either voltage or currently crosses the zero axis, the power wave must also cross the zero axis. (Zero times any number is zero.) Inasmuch as the power curve results from multiplying instantaneous voltage

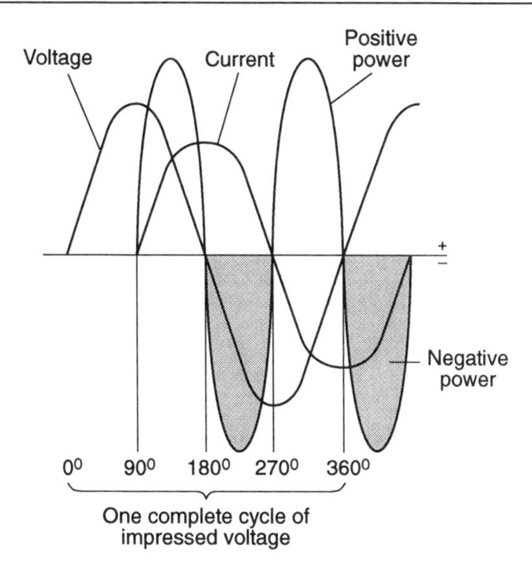

Fig. 1.5 *Voltage, current and power curves for an ideal inductance*

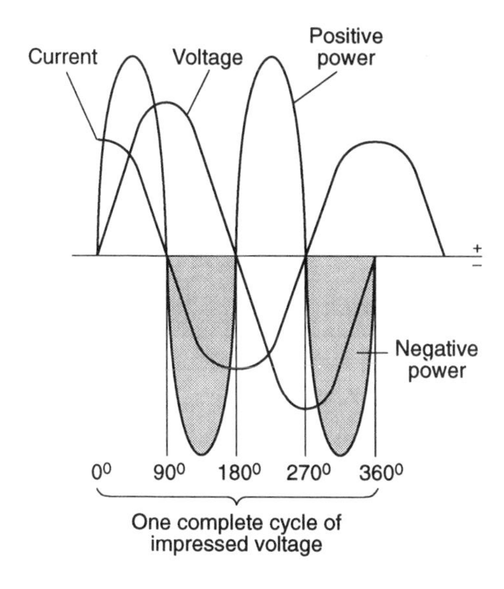

Fig. 1.6 *Voltage, current and power curves for an ideal capacitance*

by instantaneous current valves, we see why the power curve is twice the frequency of the voltage or current waves. Although negative voltage is every bit as good as positive voltage and despite the fact that the same is true for negative and positive current, this reasoning cannot be extended to

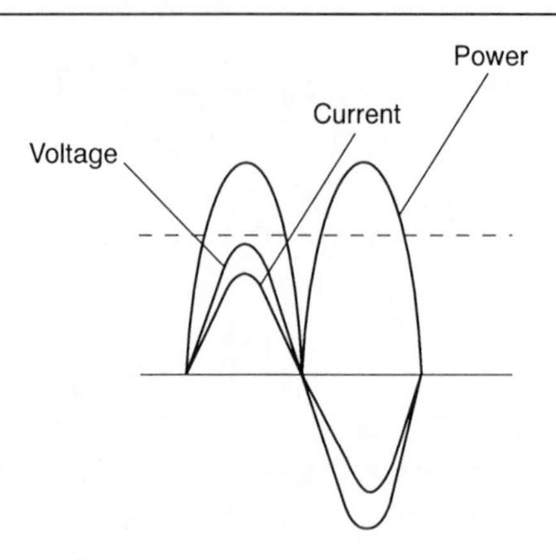

Fig. 1.7 *Voltage, current and power curves for an ideal resistance. The two loops of the power curve (the largest of the three curves) should be the same size and shape*

explain the physical significance of negative power. Positive power is the power taken from the a.c. source by the load; negative power is returned to the source from the load.

Although the ideal inductor and the ideal capacitor do not themselves consume or dissipate power, the current that they cause to flow in the line can and does cause power loss in the resistance of the line or connecting conductors and in the internal resistance of the source. We see that it still holds true that current flowing through a resistance causes power loss. Thus, although our ideal elements would produce no power loss *within themselves*, their insertion in a circuit must, nevertheless, cause power loss within other portions of the circuit. These matters are fundamental and should be the subject of considerable contemplation before going on. (We reflect that a flywheel, a rotating mass behaving in an analogous way to inductance, consumes no power from the engine. Also, an ideal, that is, frictionless, spring returns all of the mechanical power used to deflect it.)

Performance of ideal tank circuit

We are now better prepared to consider the performance of the ideal tank circuit formed by connecting the ideal inductor and ideal capacitor in parallel. Such a tank circuit is shown in Fig. 1.8. Let us suppose that the generator delivers a frequency equal to the resonant frequency of the LC combination. Resonance in such a circuit corresponds to that frequency at which the reactance of the inductor and the reactance of the capacitor are equal, but of opposite sign. From the individual properties of ideal inductors and ideal capacitors, we should anticipate that no power would be consumed from the source. This is indeed true. However, we would also find

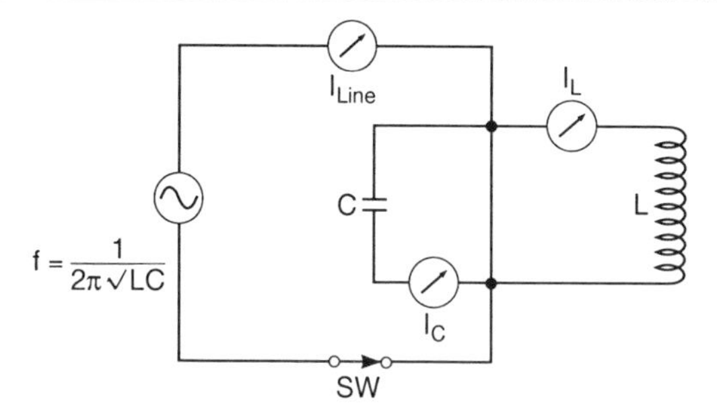

Fig. 1.8 *A resonant tank composed of ideal elements*

that, *at resonance*, no current is drawn from the line either. Can we correctly infer from such a situation that the source is *not* actually needed to sustain oscillations in the ideal tank circuit?

From a theoretical viewpoint such a conclusion is entirely valid. The switch in Fig. 1.8 could be opened and large currents would circulate in oscillatory fashion between the ideal inductor and the ideal capacitor. Our ideal tank circuit would now be self-oscillatory at its resonant frequency. We would have a sort of perpetual motion, but still not the variety attempted by inventors unversed in physical law. That is, the ideal tank circuit, though self-oscillatory, could not long supply power to a load; as soon as we *extracted* power from such a tank circuit, we would effectively introduce resistance, thereby destroying its ideal nature. An 'ideal' pendulum involving no frictional losses whatsoever would swing back and forth through eternity; however, if we attempted to harness the motion of the rod to perform mechanical work of some sort, we would dissipate its stored energy, thereby damping the amplitude of successive swings until oscillation ceased entirely.

Resonance in the parallel-tuned LC circuit

We saw in Fig. 1.5 that the current in an ideal inductor lags the applied voltage by a quarter cycle or 90°. We saw in Fig. 1.6 that the current in an ideal capacitor leads the applied voltage by a quarter cycle or 90°. Significantly, at some frequency, the current in the line feeding a parallel combination of ideal inductor and ideal capacitor will neither lead nor lag the applied voltage, but will be in phase with it. The frequency must be such that inductive reactance and capacitive reactance are equal numerically, for only then can exact cancellation of phase displacement between current and applied voltage occur. At other frequencies, the phase displacement between current and applied voltage will be less than 90° lead or lag, but cannot be zero.

We may think of inductance and capacitance as tending to cancel each other's power to cause phase shift between current and applied voltage. Inductive reactance, X_L, increases as frequency is increased ($X_L = 2\pi fL$). Capacitive reactance X_C decreases as frequency is increased: [$X_C = 1/(2\pi fC)$]. Therefore, at the one frequency known as the *resonant frequency*, and at this frequency only, the two reactances are numerically equal. (see Fig. 1.9). Inasmuch as the phase leading and lagging effects of the two reactances then cancel, the tank circuit no longer behaves as a reactance, but rather as a pure resistance. In the ideal tank circuit, this resistance would be *infinite* in value; in practical tank circuits, we shall find that this resistance, R_0, has a finite value dependent upon the inductance, capacitance, and resistance in the tank circuit.

Inductance/capacitance relationships for resonance

At resonance, X_L and X_C are equal. Therefore we may equate their equivalents thus, $2\pi f_0 L = 1/(2\pi f_0 C)$. If we next make the algebraic transpositions necessary to solve this identity for the resonant frequency, f_0, we obtain $f_0 = 1/(2\pi\sqrt{LC})$. We see from this equation that for a given resonant frequency, f_0, *many* different combinations of L and C are possible; we may resonate with large inductance and small capacitance, or with the converse arrangement (see Fig. 1.10). Above and below the frequency of resonance, X_L and X_C no longer cancel and the tank circuit behaves as a reactance. Specifically, at higher than resonant frequencies, the tank 'looks' capacitive; at lower than resonant frequencies the tank displays the characteristics of an inductor. If for the moment we disregard the nature of the tank circuit impedance, paying no attention to whether it is inductive, resistive, or capacitive, we can more easily pinpoint one of the most important of all properties possessed by the parallel tank circuit.

We refer here to the fact that the impedance is maximum at resonance and decreases as the applied frequency departs on either side of resonance. This means that the tank circuit will develop maximum voltage at the resonant frequency; voltages at other frequencies will be suppressed or rejected by the 'shorting' action of the relatively low impedance they experience across the tank circuit terminals. In the ideal tank circuit, the resonant frequency is supported freely across a resistive impedance infinitely high; frequencies only slightly higher or lower than the resonant frequency experience a short circuit and are thereby rejected. Practical tank circuits gradually discriminate against off-resonant frequencies (see Fig. 1.11). The better they do this, that is, the greater the rejection for a given percentage of departure from true resonance, the greater we saw is the selectivity of the tank circuit. In Fig. 1.11, the resistance R is not part of the tank circuit, but is used to isolate the tank circuit from the source. Otherwise, a low impedance source could mask the impedance variations of the tank circuit.

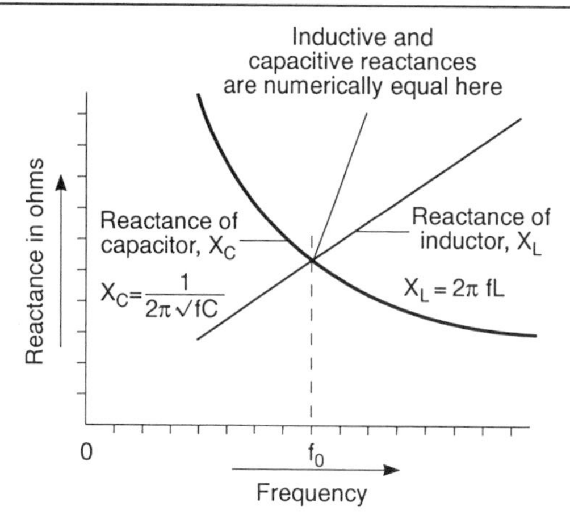

Fig. 1.9 *Variation of inductive and capacitive reactance with frequency*

$$f_0 = \frac{1}{2\pi\sqrt{LC}}$$

$$= \frac{1}{2\pi\sqrt{(50 \times 10^6 \times 500 \times 10^{-12})}}$$

$= 1.0$ MHz

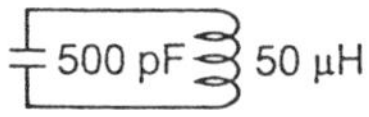

500 pF 50 μH

$$= \frac{1}{2\pi\sqrt{(500 \times 10^6 \times 50 \times 10^{-12})}}$$

$= 1.0$ MHz

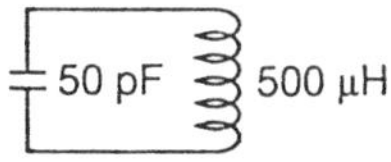

50 pF 500 μH

$$= \frac{1}{2\pi\sqrt{(500 \times 10^6 \times 5 \times 10^{-12})}}$$

$= 1.0$ MHz

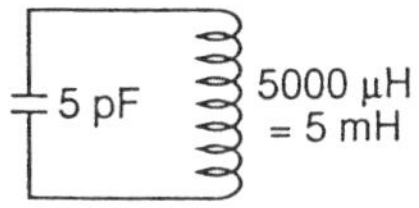

5 pF 5000 μH = 5 mH

Fig. 1.10 *Different combinations of* L *and* C *for obtaining the same resonant frequency*

The following conditions at resonance of a parallel LC tank circuit (R_L and R_C are small) apply in Fig. 1.12:

- the applied voltage, V, and the line current, I_{Line}, are in phase
- inductor current, I_L, and capacitor current, I_C are of equal value because inductive reactance and capacitive reactance are equal
- the voltage, V, applied across the tank circuit rises to its maximum value because the impedance of the tank circuit attains its highest value
- the tank circuit behaves as a pure resistance
- the line current, I_{Line}, attains its minimum value

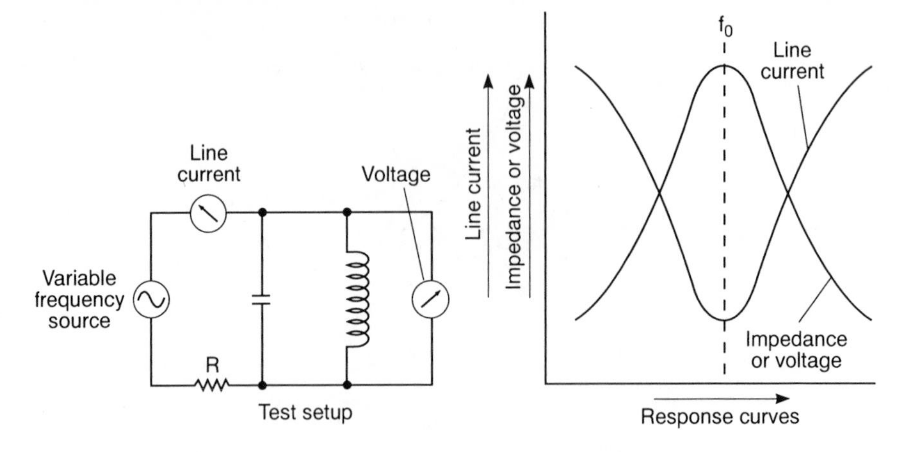

Fig. 1.11 *Typical frequency response of a parallel LC circuit*

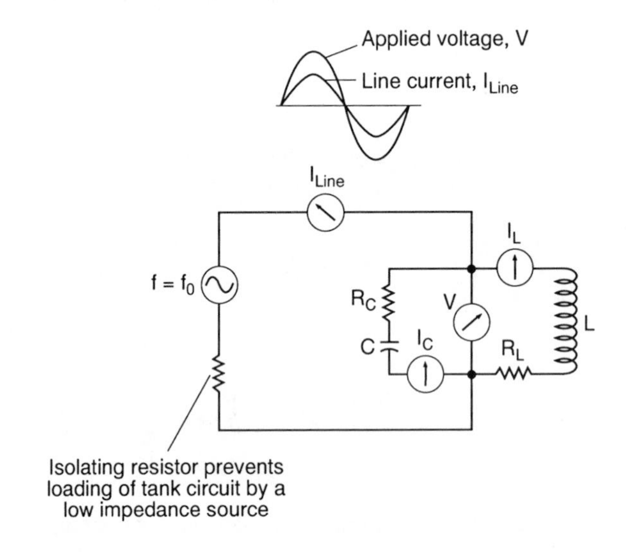

Fig. 1.12 *Resonance conditions in a typical LC tank circuit*

Practical tank circuits with finite losses

We have been dealing with *ideal* reactive elements and *ideal* tank circuits in order to increase our insight into the nature of actual tank circuits. We realize that practical inductors and capacitors contain losses. Let us now investigate the effect of these losses in the actual tank circuit. First, we can no longer expect zero line current at resonance. Although the line current will be minimum at resonance, a residual value will exist. This must imply that

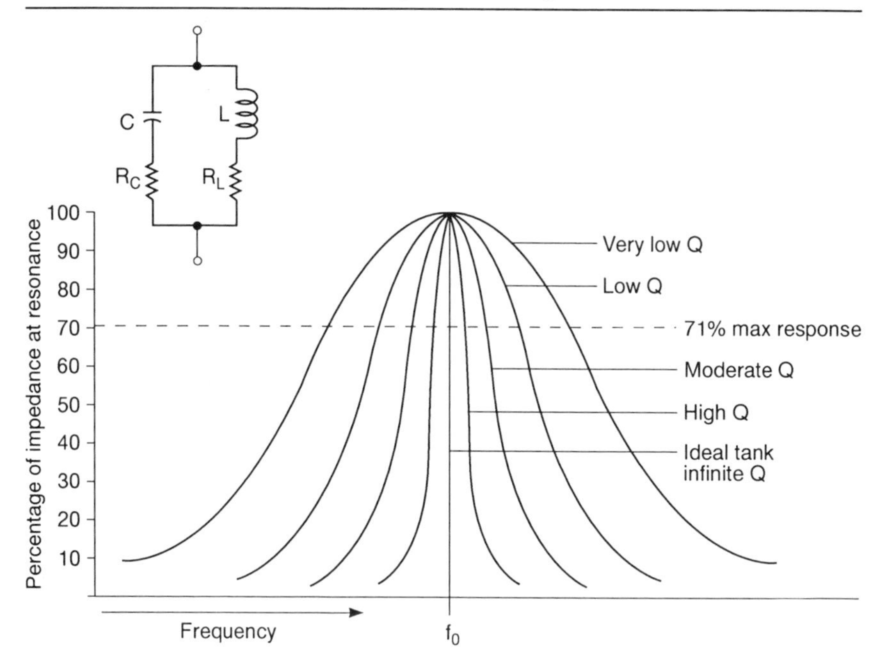

Fig. 1.13 *Resonance curves showing effect of Q on frequency selectivity*

the terminal impedance at resonance is no longer of infinite value (see Fig. 1.12). In order to acquire a working acquaintance with tank circuits, it is expedient that we learn how the several relevant properties of the tank circuit are interrelated. To make such a study meaningful, we must understand the significance of the figure of merit known as the '*Q*' of a tank circuit.

There are several different ways to think of *Q*. However, notwithstanding the definition best suited to a particular tank circuit application or calculation, *Q* is basically a measure of the ratio of energy stored per cycle to the energy dissipated per cycle. We see that the *Q* of the ideal tank circuit, which contains *no* power dissipating resistance or other losses, is infinite. As a corollary of this basic concept of *Q*, we can also say that *Q* provides a measure of the frequency selectivity, that is, the bandwidth of the tank circuit (see Fig. 1.13). *Q* enables us to judge the ability of a tank circuit to suppress harmonics. From the mathematical viewpoint, *Q* constitutes an extremely convenient means of tying together the various electrical parameters of the tank circuit. In this book, we shall assume that *Q* always has a value of ten or more. When tank circuit losses are great enough to reduce *Q* below this value, the simple formulas for resonance and for other properties of the tank circuit are no longer reasonably accurate and the resonance curve loses its symmetry. In Fig. 1.13, when the sum of R_C and R_L is low, the *Q* is

1. $Q_0 = \dfrac{R_0}{(2\pi f_0)(L)}$ — Where R_0 is the impedance across the tank terminals at resonant frequency f_0.
2. $Q_0 = (R_0)(2\pi f_0)(C)$
3. $Q_0 = \dfrac{1}{(2\pi f_0)(C)(R_S)}$ — Where R_S is the sum of the losses in the inductor and capacitor expressed as equivalent series resistances. That is, $R_S = R_L + R_S$ in Fig 1.12. In many practical cases, we may reasonably say that $R_S = R_L$, R_C being negligible.
4. Thus, $Q_0 = \dfrac{1}{(2\pi f_0)(C)(R_L)}$ — Where R_C is negligible.
5. $Q_0 = \dfrac{(2\pi f_0)(L)}{R_L}$ — It is aso true that $Q = \dfrac{(2\pi f)(L)}{R_L}$, that is, the Q of the inductor at any frequency, f, is obtained by substituting f for f_0. (Inasmuch as losses vary with frequency, R_L is likely to be a different value if f differs appreciably from f_0.)
6. $Q_0 = \dfrac{(0.5)(f_0)}{f_0 - f_x}$ — Where f_x is the frequency below or above resonance which causes the impedance (or voltage) across the tank terminals to be 70% of the impedance (or voltage) at resonance.
7. $Q_0 = \dfrac{(2\pi)(\text{energy stored per cycle})}{(\text{energy dissipated per cycle})}$ or, for practical purposes,

   $Q_0 = \dfrac{\text{volt-amperes}}{\text{watts}}$ which is obtained from $\dfrac{(V)(I_{Line})}{W}$
8. Also, $Q_0 = \dfrac{I_L}{I_{Line}} = \dfrac{I_C}{I_{Line}}$ when the tank is at resonance.
9. In Q_0/π cycles the oscillating voltage across a shock-excited tank circuit falls to 37% of its maximum value if additional disturbances are not applied to the tank circuit.
10. In $Q_0/2\pi$ cycles, the stored energy in a tank circuit falls to 37% of its maximum value unless replenished before elapse of that time. We imply here, the time interval following a pulse or disturbance, or after diconnection of the a.c. source.
11. Significance of $Q_0 = \frac{1}{2}$ or less — When sufficient resistance is introduced to make $Q_0 = \frac{1}{2}$, a resonant circuit is critically damped; no oscillation can be provoked by transients or shock excitation. This condition prevails when the tank circuit is overdamped by making Q_0 less than $\frac{1}{2}$.

Chart 1.1 *Relationships of the Resonant Q Factor (Q_0) in the Parallel LC Tank*

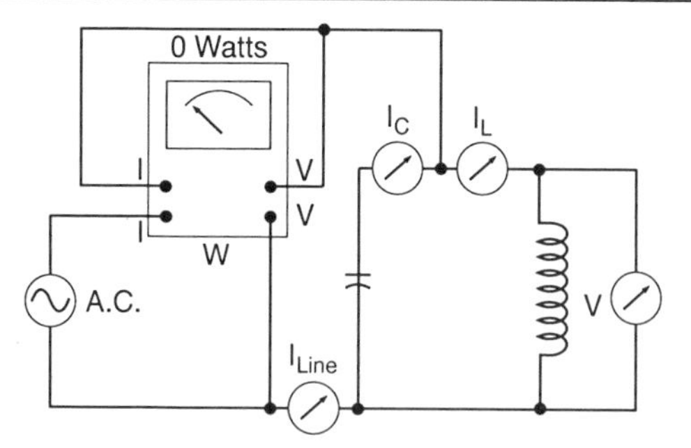

Fig. 1.14 *Parallel LC tank circuit for use with Charts 1.1, 1.2 and 1.3*

high. When the sum of R_C and R_L is high, the Q is low. It fortunately happens that the vast majority of LC networks intended to function as tank circuits have Qs that exceed ten. Qs of 20 to 100 are commonly found in receivers. Qs of several hundred may be found in transmitters. Quartz crystals and microwave cavities commonly have Qs of several tens of thousands and the Q of special quartz crystals may be on the order of a half million.

Figure of merit, 'Q'

We will concern ourselves primarily with the value of Q that exists at (or very close to) resonance. It will be convenient to designate Q for this condition as Q_0. A great deal of insight into the parallel LC tank circuit can be gained by a serious study of the Q_0 relationships shown in Chart 1.1 (also see Fig. 1.14). The fact that Q_0 can be expressed in so many ways is, in itself, indicative of the importance of this tank-circuit parameter. We see that a knowledge of Q and one quantity other than f_0 is sufficient to enable us to start a chain of calculations from which we can determine *all* of the many parameters associated with the parallel LC tank circuit.

Physical interpretation of R_0

R_0 is the impedance seen across the tank terminals at resonance. Significantly, R_0 appears as a pure resistance. It is interesting to contemplate that, although R_0 is purely resistive, its value is largely governed by *reactances*. This is revealed by the two relationships:

$$R_0 = Q_0\ (2\pi f_0)(L)$$

and

1. $R_0 = (Q_0)(2\pi f_0)(L)$
2. $R_0 = \dfrac{Q_0}{(2\pi f_0)(C)}$
3. $L = \dfrac{R_0}{(Q_0)(2\pi f_0)}$
4. $C = \dfrac{1}{(Q_0)(R_S)(2\pi f_0)}$
5. $C = \dfrac{1}{(Q_0)(R_L)(2\pi f_0)}$
6. $R_S = \dfrac{(2\pi f_0)(L)}{Q_0}$
7. $L = \dfrac{Q_0(R_S)}{2\pi f_0}$
8. $R_L = \dfrac{(2\pi f_0)(L)^*}{Q_0}$
9. $L = \dfrac{(Q_0)(R_L)^*}{2\pi f_0}$
10. $I_L = I_L = (Q_0)(I_{Line})$

*When capacitor losses are negligible.

Chart 1.2 *Calculation of Parallel LC Tank Circuit Parameters with the Aid of the Resonant Q Factor, Q_0*

1. $X_L = X_C = \sqrt{\dfrac{L}{C}}$ — This relationship holds true only at resonance.
2. $BW = \dfrac{1}{(2\pi)(R_0)(C)}$ — These formulas enabled us to calculate the bandwidth of a parallel LC tank circuit. *BW* is the bandwidth in hertz where the impedance, or voltage, is 71% of the value at resonance.
3. $BW = \dfrac{R_S}{(2\pi)(L)}$
4. $X_L = (2\pi f)(L)$ — Where X_L is the reactance of the inductor at any frequency, *f*.
5. $X_C = \dfrac{1}{(2\pi f)(C)}$ — Where X_C is the reactance of the capacitor at any frequency, *f*.
6. $f_0 = \dfrac{1}{2\pi\sqrt{LC}}$
7. $R_0 = \dfrac{L}{(R_S)(C)}$

Chart 1.3 *Additional Relationships Existing in the Parallel LC Tank Circuit*

$$R_0 = \frac{Q_0}{(2\pi f_0)(C)}$$

In these relationships, we note that $(2\pi f_0)(L)$ is inductive reactance and $1/(2\pi f_0)(C)$ is capacitive reactance, both reactances being designated for the resonant frequency f_0.

Chart 1.2 is included to depict the numerous parallel LC tank calculations that involve the resonant value of Q, that is, Q_0. Chart 1.3 is a list of other

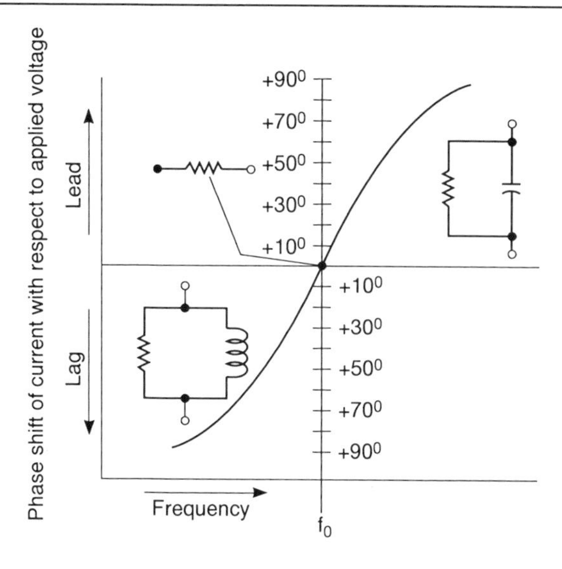

Fig. 1.15 *Typical phase response of a parallel LC tank circuit*

useful formulas relating various parameters of the parallel LC tank circuit. In these three tables, resistances such as R_0, R_L, R_C and R_S must be expressed in ohms. Reactances, such as X_L and X_C, must also be expressed in ohms. Frequencies such as f, f_0 or f_x must be expressed in hertz. The same is true of bandwidth, BW. Inductance, L, must be expressed in henrys, and capacitance in farads. Finally, Q and Q_0 are dimensionless numbers designating the figure of merit of the tank circuit.

Phase characteristics of parallel-tuned LC circuit

Before we leave the parallel LC tank circuit, we should visualize its effect upon the phase relationship between impressed voltage and resulting line current. Figure 1.15 illustrates the general nature of phase response in such a circuit. Significantly, we see that at resonance, there is no phase shift. Below the resonant frequency, the circuit acts as an inductor in parallel with a resistance. Consequently the tank 'looks' inductive. The further we depart from resonance in the frequency region below resonance, the greater the effect of the inductance and the less the effect of the resistance. (The effective parallel resistance is of the same nature as R_0 but at different frequency.) At low frequencies, the phase shift approaches 90° and the tank appears therefore as a more nearly perfect inductor. Above the resonant frequency, the circuit acts as a capacitor in parallel with a resistance. (The resistance is again of the same nature as R_0, but corresponds to a frequency higher than f_0.) At high frequencies, the phase shift approaches 90°, but in the opposite direction as in the low frequency case. The tank appears therefore as a more nearly perfect capacitor.

Series-resonant tank circuits

In oscillators, and in the equivalent circuits of oscillating elements, we often encounter the *series-resonant* combination of inductance and capacitance, and, of course, the inevitable resistance. Such a configuration also behaves as a 'tank', that is, its properties are the manifestations of energy storage. The characteristics displayed by the series-tuned tank circuit are, generally speaking, opposite to those of the parallel-tuned tank circuit. In the parallel tank circuit, we found that the line current attained its minimum value at resonance. Conversely, in the series tank circuit, line current is maximum at resonance. In a series tank circuit composed of a lossless inductor and a lossless capacitor, the line current would reveal the presence of a perfect short circuit placed across the a.c. source. In practical series tank circuits, line current at resonance is limited only by the equivalent series resistances of the inductor and capacitor, which are, respectively, R_L and R_C.

Q in series tank circuits

Our concepts of Q and Q_0 are in principle still valid for the series tank circuit; however, we must now use a different set of formulas to relate the Q valve to the various parameters of the series tank circuit. The important relationships in the series tank circuit are given in Chart 1.4. As previously was the case, we must stipulate a minimum Q_0 of ten in order to preserve the validity of these formulas. In most applications encountered in practice, this will not restrict our ability to visualize the operation of the series tank or prevent us from calculating or estimating reasonably accurate parameter values, for a Q_0 below ten is the exception rather than the rule.

Resonance in series-tuned LC circuit

We observe that the resonance curve is essentially of the same nature as in the parallel tank; however, we now deal with *line current* rather than applied voltage (or impedance) as a function of frequency (see Fig. 1.16). At the resonance frequency, f_0, the line current is restricted by the equivalent series resistance, R_S, representing the losses in the inductor, R_L, and the losses in the capacitor R_C. When we inspect the phase response of the series tank in Fig. 1.17, we see that it is essentially the converse of the phase response

1. $Q_0 = \dfrac{(2\pi f_0)(L)}{R_S}$
2. $Q_0 = \dfrac{(2\pi f_0)(L)}{R_L}$ — Where capacitor losses R_C are much less than inductor losses R_L
3. At resonance $V_L = V_0 = Q_0 V_{Line}$
4. $R_r = \dfrac{V_0}{I_0}$ — Where R_r is the effective resistance of the circuit seen by the a.c. source when resonance exists. R_r in the series tank corresponds to R_0 in the parallel tank. Where V_0 is the terminal voltage at resonance. Where I_0 is the line current at resonance.
5. $R_r = R_S = R_L + R_C$
6. $Q_0 = \dfrac{L}{(2\pi f_0)(C)(R_S)} = \dfrac{L}{(2\pi f_0)(C)(R_L)}$ — When R_C is negligible.
7. $I_0 = \dfrac{V_0}{R_S}$

Chart 1.4 *Parameters of the Series Tank Circuit*

associated with the parallel tank circuit shown in Fig. 1.15. Whereas the parallel tank circuit is inductive at frequencies below resonance, we find the series circuit is capacitive in this frequency region. Similarly, at frequencies above resonance, the series circuit is inductive rather than capacitive as we found to be the case in the parallel tank circuit. It is interesting to note that the voltages existing across the inductor and capacitor of the series circuit can be many times the value of the line voltage impressed across the circuit terminals (see Fig. 1.18). We recall the counterpart of this phenomenon in the parallel tank circuit wherein the circulating current, i.e., I_L and I_C, can be many times the value of the line current, I_{Line}.

L/C ratio in tank circuits

Of considerable practical importance is the effect of the inductance-to-capacitance ratio on Q_0 in tank circuits. In both parallel and series tank circuits, a given resonant frequency, f_0, can be attained by numerous combinations of inductance and capacity (see Fig. 1.10). In the parallel tank circuit, for a given value of R_0 and f_0, we increase Q_0 as we make the capacitance larger and the inductance smaller. However, in the series tank circuit, Q_0 is increased, for a given value of R_S and f_0, by making the capacitance smaller and the inductance larger. Both the parallel tank circuit and the series tank circuit

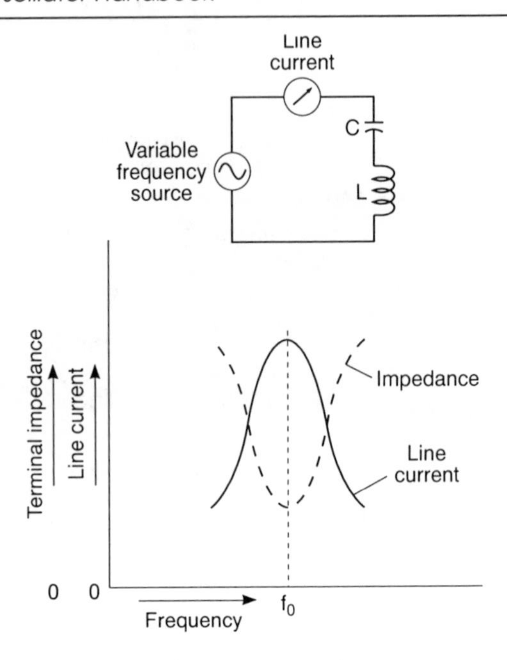

Fig. 1.16 *Typical frequency response of a series LC tank circuit*

are prevented from having an infinite Q_0 by the presence of R_S. Whereas high capacitance is necessary to make Q_0 high in the parallel tank circuit, high inductance is necessary to make Q_0 high in the series tank circuit. These statements are mathematically true, but in practice we cannot increase either the capacitance in the parallel tank, or the inductance in the series tank indefinitely without incurring losses, which ultimately bring us to a point of diminishing returns, where in Q_0 begins to decrease rather than increase.

For example, any resonant frequency, f_0, can be attained in a parallel tank by means of a one-turn coil and an appropriate capacitor. However, as we make f_0 lower in frequency, the capacitor required to achieve resonance becomes larger. The dielectric losses in a physically large capacitor would mount up until further expansion of its physical size would add sufficient losses to cause subsequent decrease of Q_0 with any further pursuit of this trend.

A similar argument can be presented for the inductor in the series tank. Thus, in practical tank circuits of both varieties, we find that we are restricted to a Q_0 value, which cannot be exceeded by further change in the ratio of inductance to capacitance. Furthermore, in the parallel tank circuit, the change in the value of R_0 with the LC ratio introduces other difficulties. Often, the parallel tank is required to have an R_0 value within a certain range. This may conflict with the Q_0 requirement and a compromise must

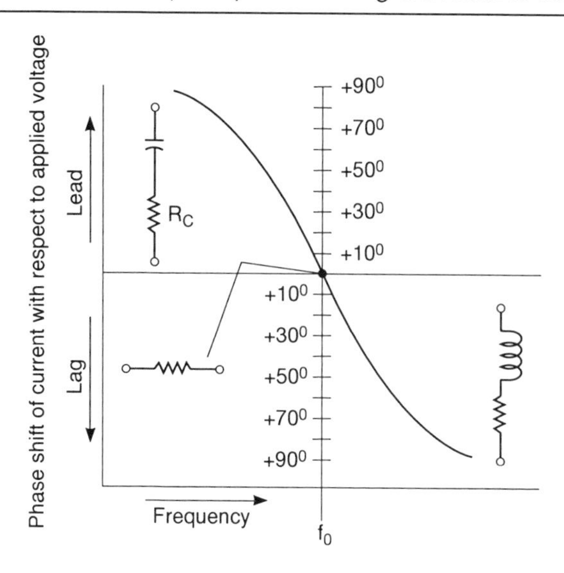

Fig. 1.17 *Typical phase response of a series-tuned LC circuit*

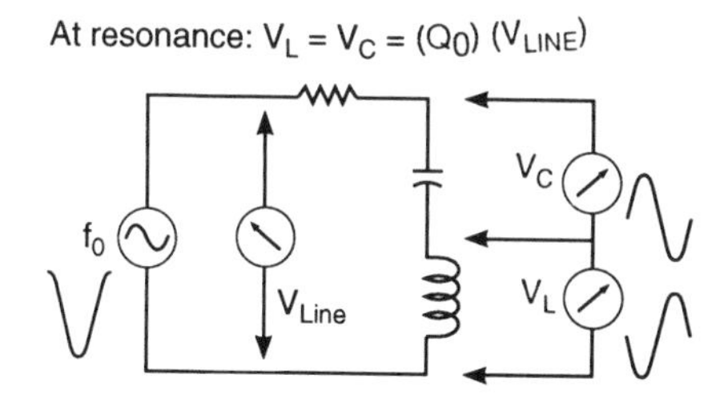

Fig. 1.18 *Voltage step-up in a resonant series tank circuit*

be made. Also, we should appreciate the fact that an extremely high Q_0 is not always desirable. The selectivity of a high Q_0 tank circuit can be sufficient to discriminate against the frequency spectrum required in voice modulation.

In other instances, the sharp tuning of the high Q tank may result in critical adjustments that are not mechanically stable. In transmitters, diathermy, and induction and dielectric heating equipment, too high a Q produces excessive circulating current in parallel tank circuits, thereby dissipating power and generating heat where least desired. Finally, we note that in both types of tank circuits, it is possible to have losses and a high Q_0 simultaneously. For example, series-mode resonance in a quartz crystal is accompanied by a relatively high value of R_s. However, Q_0 is nevertheless very high due to the high *ratio* of effective L to R_s.

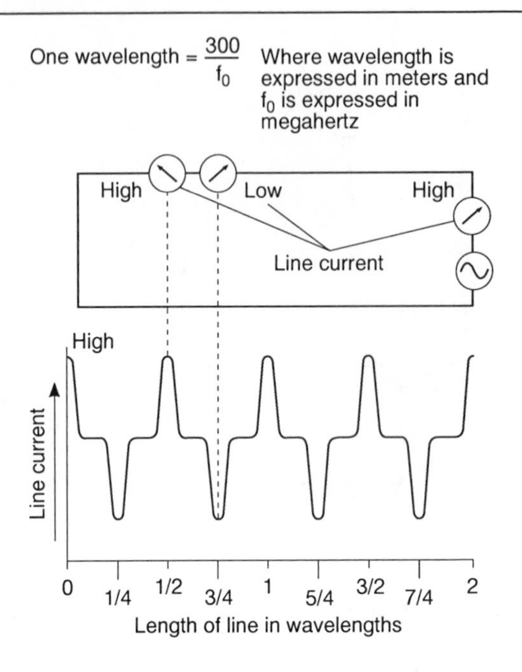

Fig. 1.19 *Parallel- and series-type resonances in a short-circuited line*

Transmission lines

Transmission lines, such as parallel wires or coaxial cable, are capable of producing resonance phenomena similar to that associated with 'lumped' inductance and capacitance elements. The inductance and capacitance are present in the transmission line, as in the coil and capacitor, but are distributed along the length of the line. If the end of the line farthest from the generator is either open-circuited or short-circuited, resonances will exist at distances corresponding to quarter-wavelength spacings from that end. From the viewpoint of the generator, that is, in terms of the line current it is called upon to deliver, the line alternates from series to parallel resonance each time we change the line length one quarter wavelength. This is shown in Fig. 1.19 in another manner.

We see that series-type and parallel-type resonances occur at successive quarter-wave distances from the shorted end of the line. An open line of any number of odd wavelengths long behaves similarly, that is, it undergoes a similar type of resonance, to a shorted line of any number of even wavelengths long. For our study of oscillators, we shall be primarily interested in parallel resonance of transmission line. In Fig. 1.20 we see the simplest line configurations for simulating parallel resonance. At the terminals of the

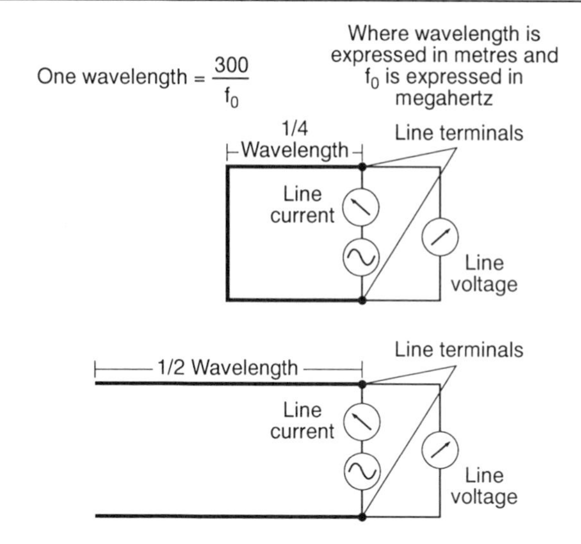

Fig. 1.20 *Simplest resonant lines for simulating the properties of the parallel-tuned tank circuit*

two configurations, voltage is maximum and current is minimum, as at the terminals of a parallel LC tank circuit.

In speaking of a resonant frequency of lumped tank circuits, we found ourselves limited to exactly *one* such frequency for any given LC combination. However, a transmission line can exhibit resonances at many frequencies corresponding to various multiplies of a quarter wavelength. For example, a line that is shorted at its far end, and that is a quarter wavelength at f_1 displays series and parallel-type resonances respectively at $2f_1$, $3f_1$, $4f_1$, $5f_1$, etc. The series resonances do not cause oscillation in oscillators designed to use the properties of parallel resonance. The parallel resonances can result in oscillation at other than the desired frequency, but in practice, it is generally not difficult to ensure oscillation at only the intended frequency. One of the factors that tends to discourage oscillation at such frequencies as $3f_1$, $5f_1$, $7f_1$, etc., in the shorted line (or at $2f_1$, $4f_1$, $6f_1$, etc., in the open line) is that the losses become higher at the multiple frequencies of f_1. Other things being equal, an oscillator will 'prefer' to operate at the frequency involving lowest losses, i.e., highest tank circuit Q.

The delay line

The delay line is an important element for certain oscillators of the relaxation type. In these applications, *pulse duration*, rather than pulse repetition rate or frequency, is the controlled parameter. Delay-line stabilized oscil-

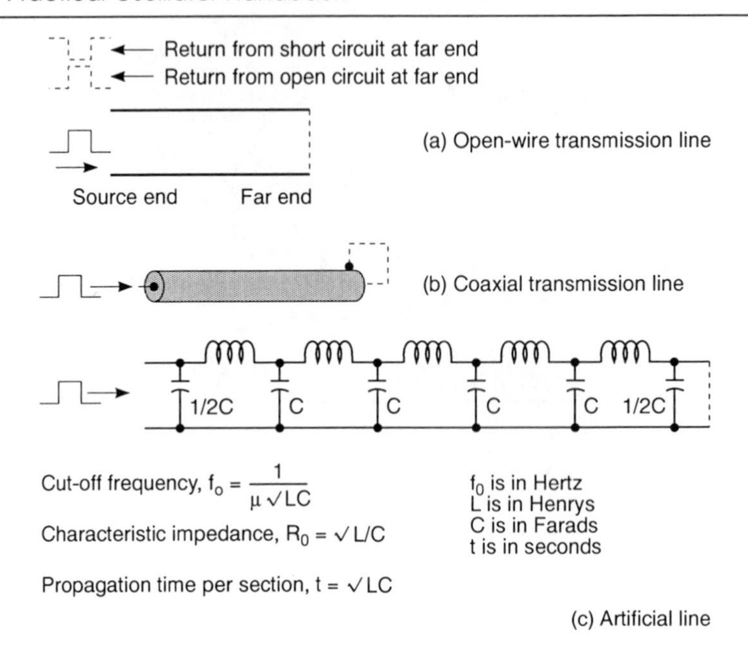

Fig. 1.21 *Delay lines*

lators are used quite extensively in radar and in digital techniques. The delay line is essentially a transmission line. Nevertheless, its operating mode, and generally its physical appearance also, differ from the lines employed as high-Q resonating elements in sinusoidal oscillators.

The simplest delay line comprises a length of open wire or coaxial transmission line. Pulses introduced at one end require a precise time interval to reach the opposite end. If the far end of the line is terminated in a resistance equal to the characteristic impedance of the line, the pulse dissipates its energy in the resistance. If the far end of the line is either open circuited or short circuited, the pulse undergoes reflection and is thereby returned to the source end of the line. Under such conditions, a more efficient use is made of the lines because the pulse is forced to traverse its length *twice*. Thus, the obtainable time delay is twice what it would be if the pulse energy was absorbed in a resistor at the far end of the line (see Fig. 1.21).

If a line is open circuited at its far end, the returned voltage pulse will have the *same* polarity as the initiating pulse. If the line is short circuited at its far end, the returned voltage pulse will have *opposite* polarity with respect to the initiating pulse voltage. The latter case is particularly useful, for the arrival of the reflected pulse is readily utilized to *terminate* the conduction state of the oscillator responsible for the generation of the leading edge of the pulse. In this way a precise pulse duration is established, being determined by the *electrical length of the line* rather than by time constants of resistances and capacitances associated with the oscillator. Variations in power supply voltage and tube or transistor temperature cannot exert appreciable effect upon pulse duration when an oscillator is stabilized by a delay line.

An electrical disturbance propagates along typical transmission lines at speeds between 60% and and 98% the speed of light in free space, depending upon the dielectric employed for line spacing. A delay of one microsecond would require a line length on the order of a city block! Thus, except when delays of a small fraction of a microsecond suffice, the ordinary transmission line would assume impractical physical proportions. Fortunately, it is feasible to construct compact *networks* which simulate transmission line operation, but at relatively slow speeds of pulse propagation (long delay times).

The artificial transmission line

From electric wave filter theory, it is known that the low-pass filter configuration behaves in many respects as a transmission line. A few inductances and capacitances can be connected in such a network to provide a physically compact artificial transmission line from which relatively long delay times can be attained. Three important operating parameters are associated with the artificial transmission line. These are: t, the delay time per section (such a network generally consists of a number of cascaded elementary filter sections), the cut-off frequency, f_0, and the characteristic impedance, Z_0. All three of these parameters are governed, although in different ways, by the values of inductances and capacitances. Other things being equal, pulse fidelity tends to be better as the cut-off frequency is made *higher*, and also as more elementary, or 'prototype' sections are cascaded. For proper operation, the characteristic impedance should be nearly the same as the internal generator resistance of the oscillator during pulse production. These requirements impose contradictory design approaches and the construction of delay lines has become a competitive art for specialists. One of the criterions of performance of these devices is the *ratio of pulse delay time to pulse rise time.* The 'goodness' of a delay line increases as this ratio becomes higher.

Delay-line stabilized blocking oscillator

The repetition rate of the blocking oscillator shown in Fig. 1.22a is primarily governed by the resistance-capacitance combination, R_1R_1. The pulse duration, however, is precisely determined by the delay line. The electrical length of the delay line corresponds to a shorter period of time than the pulse would endure in the absence of the delay line. When the blocking oscillator commences its 'on' state, the emitter of the pnp transistor becomes positive with respect to its base (and ground). This initiates propagation of a positive-going wavefront down the line. Reflection occurs at the shorted (grounded)

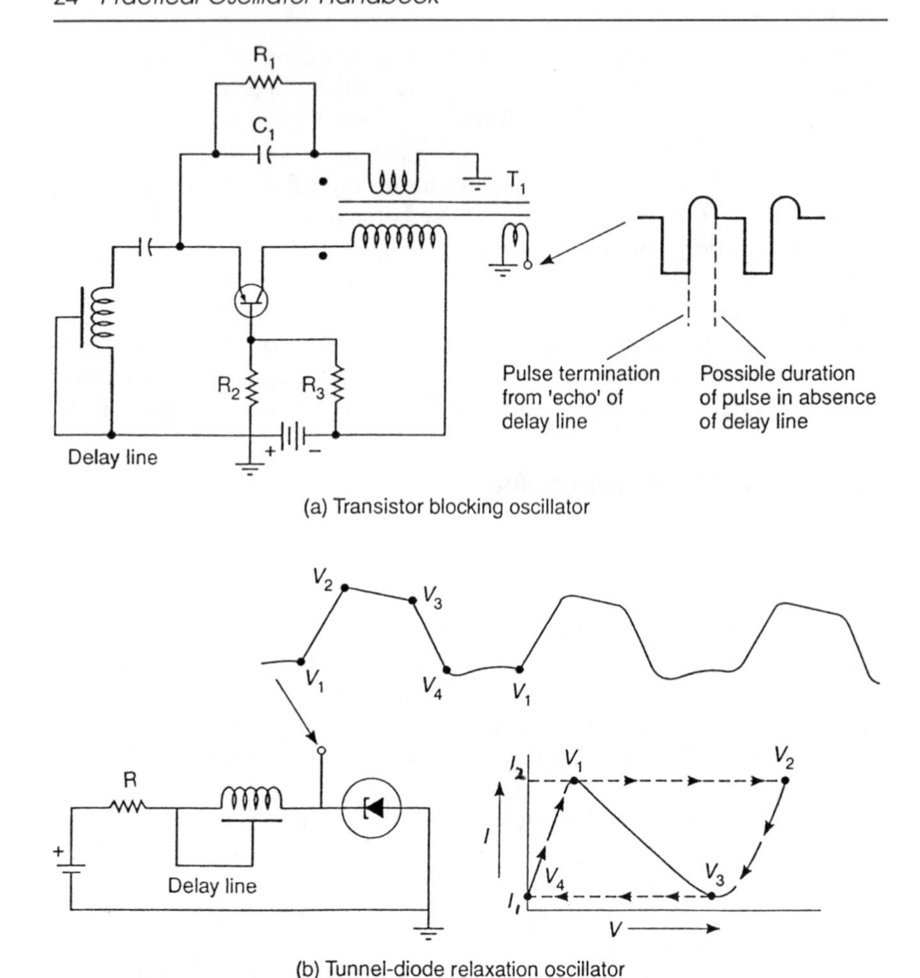

Fig. 1.22 *Stabilization of wave duration by use of delay lines*

end of the line and an inverted voltage pulse is propagated back toward the emitter end of the line. When the wavefront of this reflected pulse arrives at the emitter, the transistor is deprived of forward conduction bias. This causes the blocking oscillator to abuptly switch to its 'off' state, thereby terminating the duration of its generated pulse.

Delay-line stabilized tunnel-diode oscillator

In Fig. 1.22b, a simple but very useful oscillator is shown. Relaxation oscillations occur, but with the pulse duration controlled by the shorted delay line. It will be noted that the delay line is used in place of the inductor

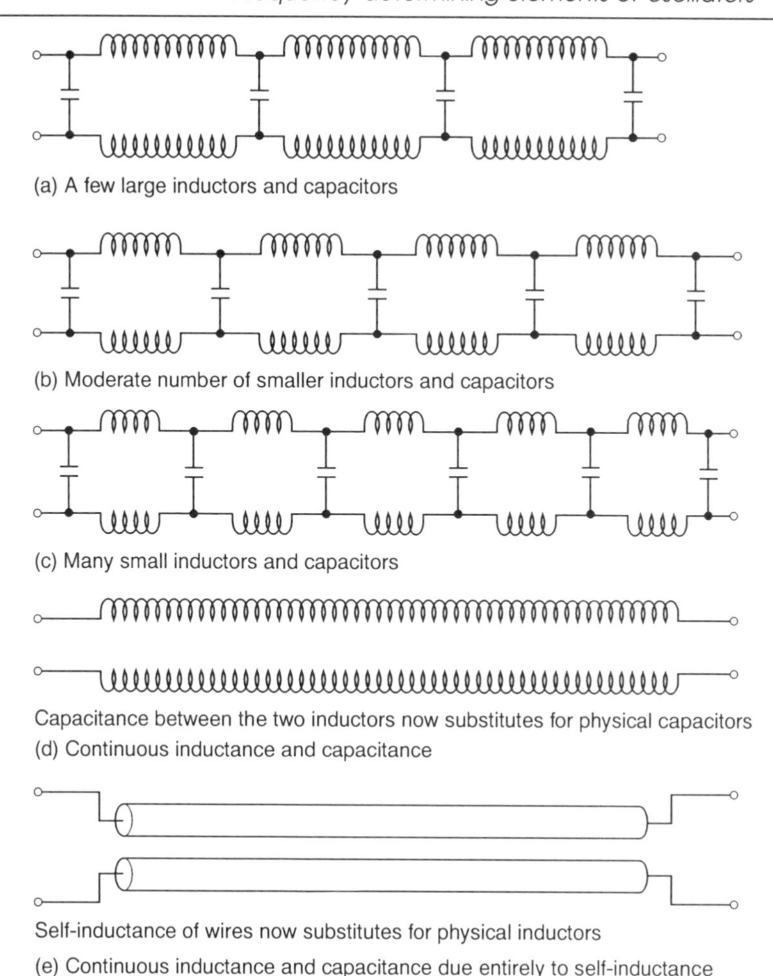

Fig. 1.23 *Synthesis of a resonant transmission line from lumped elements*

ordinarily associated with this circuit. The value of R must be high enough to permit switching transitions to occur between characteristic curve points shown in the accompanying illustrations.

Distributed parameters from 'lumped' LC circuit

In order to bridge mentally the transition between 'lumped' tank circuits and resonant lines, a somewhat different conceptual approach is needed.

This may be acquired with the aid of Fig. 1.23, which shows how a resonant line can be theoretically evolved from lumped elements. At Fig. 1.23a, several relatively large inductors and capacitors are connected to simulate roughly the condition in a resonant line. An a.c. generator connected to the left pair of terminals will deliver a current which will undergo cyclic maxima and minima as the frequency is varied. Maximum current, from the viewpoint of the generator, corresponds to series resonance; conversely, minimum current corresponds to the resonance associated with the simple parallel-tuned LC tank. For either type of resonance, voltages and currents will now be distributed throughout the network, rather than concentrated at single circuitry junctions as in a simple tank circuit.

For example, the voltage across alternate capacitors will be very nearly the same. We have simulated the alternate series and parallel type resonance displayed at quarter-wave intervals along a line. At Fig. 1.23b we reduce the size of the inductors and capacitors, but employ more of them. At Fig. 1.23c, this process is carried still further. At Fig. 1.23d we arrive at a continuous inductance and no longer rely for inter-line capacitance on physical capacitors. The required capacitors are now so small and are spaced so close together that the inherent capacitance existing *between* the 'lines' serves the required function. Finally, at Fig. 1.23e, we stretch out the helical coil and rely upon the self-inductance of the straight wires to contribute the necessary inductance.

Resonance in transmission lines

We will be primarily interested in the resonant conditions of lines. These are most effectively displayed by lines that are either short-circuited or open-circuited at their far ('receiving') ends. However, additional insight into the nature of such lines is attainable by comparing them further to lumped tank circuits. One such comparison involves termination by a resistance value that prevents resonances. The value of such a terminating resistance is given by the formula:

$$R = \sqrt{\frac{L}{C}}$$

where:

R is the resistance in ohms,
L is the inductance in henrys per unit length of line,
C is the capacitance in farads per unit length of line.

Interestingly, a similar situation prevails for lumped tanks. In Fig. 1.24 we see that the destruction of resonance in both instances is brought about by resistance values equal to $-\sqrt{L/C}$. In the line, this value is called the

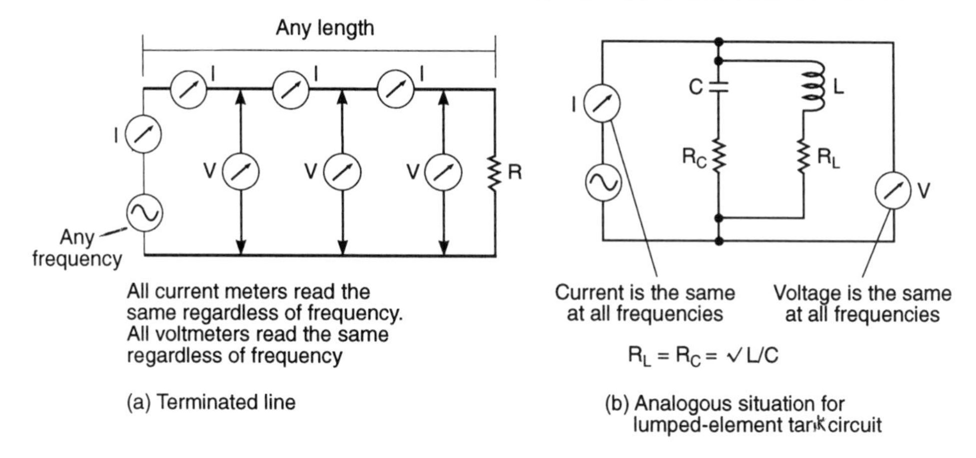

Fig. 1.24 *Termination of line by resistance equal to its characteristic impedance destroys resonance completely*

characteristic impedance of the line. The characteristic impedance of the line is determined by wire diameter, wire spacing and the dielectric constant of the insulating material. Coaxial lines operate in the same way as open lines but have the advantage that the external surface of the outer conductor may be operated at ground potential. This produces effective shielding that prevents radiation. The Q is thereby increased, and interferences and feedback troubles are virtually eliminated. The highest Q in coaxial lines is attainable when the ratio of conductor diameters is 3.6 to 1. This results in a characteristic impedance of 77 Ω.

Concept of field propagation in waveguides

The evolution of 'lumped' circuitry into spatially distributed reactance, as illustrated in Fig. 1.23b, aids our visualization of the transmission line. However, the transmission line is not the ultimate configuration; we can carry the process a step further. This is best approached by considering the coaxial transmission line. Suppose we make the central conductor smaller and smaller in diameter. What must ultimately happen? We can expect an increase in characteristic impedance and perhaps a change in the losses. However, we have nothing to correlate with our experiences with 'wired' circuits to help us ascertain the result of eliminating the central conductor altogether. We might intuitively conclude that this would end the usefulness of the cable as a means of conveying high-frequency energy. Such a deduction would indict us as victims of the commonly believed concept that physical conductors are necessary to provide a means of 'go' and 'return' for high-frequency currents. Such is not actually the case.

The physicist and mathematician view the phenomenon in a somewhat differing light. They do not refute the notion that a varying magnetic and electric field accompany the flow of current in conductors; however, they do subscribe to the notion that the flow of alternating current in such a 'circuit' as a transmission line, is the result of an alternating magnetic and electric field propagated down the line, that is, between the two conductors. If this is true, why not eliminate the need for supplying the energy by ohmic contact? Rather, let us deliver energy to the line by means of a small antenna or loop so that the energy is initially supplied in the form of fields. If we can accomplish this, why bother with a 'return' conductor? Such indeed is the logic which permits us to evolve transmission lines into waveguides and cavities.

Some important features of lines and guides

In the association of guided-wave elements with oscillators, it is not necessary in many practical situations to deal directly with the rather formidable mathematics of the electromagnetic propagation of energy in these elements. A number of physical configurations of such waveguides exist; these include two-wire transmission line, coaxial cable, hollow pipes and single wires or plane surfaces. And all of these may be associated with various dielectric material other than air. Even a rod of a dielectric material may serve as a waveguide obedient to the same relationships pertaining to 'more natural' guides. Indeed, all of these are premised on the same fundamental principles—the constraint and guidance of various patterns or 'modes' of electric and magnetic fields. Although electronic practitioners are not conditioned to thinking of transmission-line voltages and currents as the *effects* of propagating fields, rather than the other way around. The truly basic way of dealing with such phenomena needs no 'go and return' conductor. From the viewpoints of physicists and mathematicians, we have been putting the cart before the horse.

This turns out to be a voluminous topic. We will limit our discussions to quarter-wave lines and guides because of their useful property as simulated parallel-resonant 'tanks'. We shall find, however, that the practical insights attained will shed light on other configurations and functions of lines and guides. One of our main objectives will be to demystify the disparity between the physical length of a section of waveguide or line and the true electrical length. Although these can be identical, they often are quite different. For example, in the technical literature, it is commonplace to depict the length of coaxial cable serving as a quarter-wave section by the dimension, $\lambda/4$, where the Greek symbol λ represents wavelength. However, were we to make the physical length of this coaxial line a quarter wavelength as 'naturally' suggested by the free-space wavelength of the involved frequency, we would be heading for trouble.

In resolving this discrepancy, we must deal with two important parameters. One is the guide wavelength. This, we must recognize as the 'true' electrical wavelength and the value we must use in sizing the physical length of the line or guide. We shall see that the guide wavelength can be either shorter than, the same as, or longer than the free-space wavelength. It depends upon the nature of the insulating (dielectric) material used, and upon the cross-sectional shape of the guide when in the form of pipes.

The other important parameter is the low-frequency cut-off of the line or guide. This implies that only frequencies above the cut-off value can be propagated within the guide. This, as we shall see, has some interesting ramifications. For the moment, it will suffice to think of hollow pipes as high-pass filters. From the mathematical point of view, the coaxial cable and the two-wire transmission lines also have low-frequency cut-offs, the value being zero frequency or d.c. That is, such lines will propagate energy at d.c. Of course, we could infer this without resort to the logic of the mathematician because of the ohmic connections to the two 'go and return' conductors. Figure 1.25 should help clarify the comparison between transmission lines and hollow waveguides.

In the ordinary way of utilizing coaxial and two-wire transmission lines, we say that they operate in their principal mode. This is defined as the field patterns within these lines that allow transport of the very lowest frequency, which as we have seen, is d.c. It is not common knowledge by otherwise knowledgeable practitioners that higher modes can also exist. And each of these higher modes, or field patterns, will be characterized by its own low-frequency cut-off, now no longer d.c. In ordinary usage these higher modes may transport no energy because of the way the generator and load are normally connected to the line. If, however, we were to inject and extract radio frequency energy by some non-ohmic means, the higher modes could be excited and utilized to propagate electromagnetic energy down the line. So-called non-ohmic techniques could make use of loops, probes or antennas, apertures or irises. Due consideration would have to be given to the geometric orientation of such injection and extraction devices, but the salient fact is that the central conductor of the coaxial cable could then be entirely dispensed with. We have thus sacrificed the capability to carry d.c., but we have made the thought transition from coaxial cable to hollow waveguide!

We now are in a position to consider a sort of 'Ohm's law' that has practical use for all types of electromagnetic waveguides. This interesting relationship enables us to easily calculate the true physical length of lines and guides that internally act as desired sections (as quarter-wave lines, for example). The form of this relationship, shown below, is particularly well adapted to many practical situations:

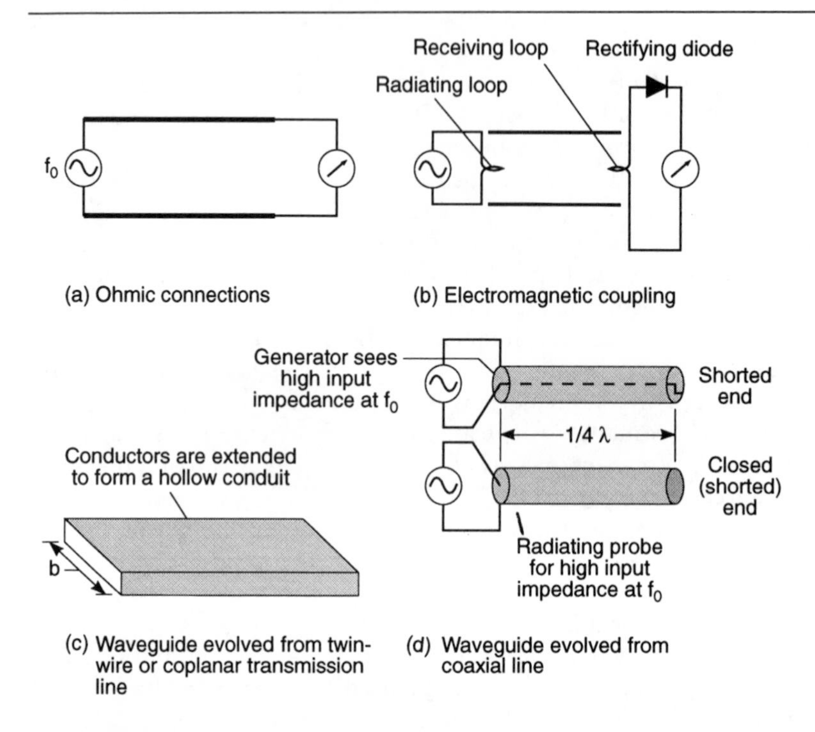

Fig. 1.25 *Evolution of hollow waveguides from transmission lines. a, Conventional use of transmission line with ohmic connections. b, Transmission line with loop-coupled injection and extraction of electomagnetic energy. D.c. and low frequencies can no longer be propagated, but higher-frequency modes can. c, Line* b *can be extended and enclosed to form a rectangular pipe. Lowest possible propagation frequency (dominant mode) now corresponds to the wavelength of twice the dimension* b. *d, A quarter wave coaxial line can dispense with its inner conductor to become a quarter-wave cylindrical waveguide*

$$\frac{\lambda_g}{\lambda_o} = \text{velocity factor} = \frac{1}{\sqrt{\varepsilon - \left(\frac{\lambda_o}{\lambda_c}\right)^2}}$$

The terms in this equation are as follows

- The velocity factor basically represents the ratio of the speed of propagation inside of the line or guide to the speed of propagation in free space. An alternative statement is that velocity factor is the ratio of the (inside) guide wavelength to the free-space wavelength. Note that the external physical length of the line or guide is theoretically the same as the guide wavelength. (In practice, we may make the external physical length a bit shorter in order to compensate for various stray reactances or because we wish to add a tuning capacitor.)

- λ_g is the guide wavelength. In coaxial lines and in two-wire lines in which most of the insulating material is air, the guide wavelength will be very nearly the same as the free-space wavelength. In such lines, a quarter-wave section will be a length of virtually the same as the free-space wavelength of the involved frequency and no calculations are necessary.
- λ_o is the free-space wavelength of the frequency being propagated. In most situations, this is a known quantity, such as the oscillation frequency of our oscillator.
- ε is the dielectric constant of the insulating material within the transmission line, or sometimes inside of the waveguide. ε is readily looked up in tables.
- λ_c is the cut-off frequency corresponding to the propagation mode we are dealing with. In both coaxial cable and two-wire transmission line, this cut-off frequency is infinite for conventional use of such lines. (Infinite wavelength denotes zero frequency or d.c.) This makes the whole fraction $\frac{\lambda_o}{\lambda_c}$ zero, and we are left with just the reciprocal of the square root of the dielectric constant.

However, if we are dealing with a pipe or rectangular waveguide, the situation is a bit different because λ_c is dependent upon shape and the cross-sectional dimensions. Because such guides are available in standardized sizes, we can consult tables. For most practical purposes, it is permissible to use $\lambda_c = 2b$, where b is the widest dimension of rectangular waveguide. For round waveguide, use $\lambda_c = 3.41r$ where r is the radius. Note that for hollow guides (air dielectric) $\varepsilon = 1$. (For both guide-shapes, these statements apply to the usually-used dominant mode.)

In order to see how the previous discussions might relate to practical situations where we might wish to associate quarter-wave lines or guides as resonators for oscillators, let us try a few examples:

Suppose a quarter-wave section of coaxial cable is desired for use in a radio frequency system operating at 25 MHz, corresponding to a free-space wavelength of about 12 m. The available coaxial cable is the type RG-8/U. The far end of the quarter-wave section will connect to the active device; the near end will be short-circuited and will be fed by the d.c. operating voltage. Neglecting strays, what should be the physical length of this line?

A quick solution consists of consulting a table such as depicted in part by Table 1.1. It is noted that the velocity factor of RG-8/U coaxial cable is 66%.

With this information, the external length of the cable section is simply $0.66 \times 12/4$ or 1.98 m. (Such a line section may, at first glance, appear

Table 1.1 *List of commonly-used coaxial cables*

Cable	Characteristic impedance (Ω)	Velocity factor (%)	Dielectric†
RG-8/M	52	75	Foam-PE
RG-8/U	52	66	PE
RG-8/U Foam	50	80	Foam-PE
RG-9A/U	51	66	PE
RG-11A/U	75	66	PE
RG-58/U	53.5	66	PE
RG-59/U	73	66	PE
RG-141/U	50	70	PTFE

† PE is solid polyethylene, Foam-PE is foamed polyethylene, PTFE is polytetrafluoroethylene, commonly known as Teflon.
Note: When the insulating material is foamed, or partially displaced by air, the dielectric constant of the material is 'diluted'. This results in a lower dielectric constant. (The dielectric constant of air is close to 1.0 and this would be accompanied by a 100% velocity factor, as is approximated in two-wire open transmission lines.)

impractical, but a convenient remedy is at hand by coiling the line. This is permissible with coax, but not generally with two-wire transmission line. Three-plus turns with a diameter of about six inches could prove suitable.) What we have done here is to calculate the internal wavelength, or λ_g, the guide wavelength. A little contemplation will show the logic to then making the external physical length equal to λ_g. (Because of inevitable stray parameters, we probably would trim the line just a bit to optimize its quarter-wave behaviour at 25 MHz.)

Interestingly, we can arrive at the same result without a velocity factor table if we know the dielectric constant of the insulation used in the RG-8/U coaxial cable. This cable utilizes polyethylene to separate the inner and outer conductors. The delectric constant of this material is close to 2.3. With this information, we can determine the velocity factor. To do this, we substitute 2.3 for ε under the square root symbol. Next, the fraction $\frac{\lambda_o}{\lambda_c}$ is scrutinized; it is noted that λ_c is infinite, corresponding to zero frequency cut-off. This effectively makes the entire fraction zero, so we are left with just the reciprocal of the square-root of 2.3. We accordingly have 1/1.517 yielding 0.659 for the velocity factor, substantially in agreement with the value previously found in Table 1.1. As before, the product of velocity factor and free-space wavelength gives us 1.98 m, the physical length of the quarter-wave line.

It should be appreciated that this second calculation reveals that coaxial line is just a special member of the family of guided-wave structures described by the mathematical equations. Note that for two-wire transmission line with air dielectric, the equation would have shown the velocity factor to be the reciprocal of square root of 1.0. Thus, such a quarter-wave line would be the length of the free-space wavelength.

Suppose now, we are dealing with a microwave system in which the oscillator is a magnetron which generates radio frequency at a wavelength of 10 cm. The microwave energy is to be piped among several function blocks and ultimately delivered to an antenna load. We wish to determine the wavelength inside of the rectangular waveguide. Such information will enable fabrication of quarter-wave sections, resonant cavities, directional couplers, slotted lines, and will be required for impedance-matching techniques. This problem is premised on the fact that, as we shall see, the guide wavelength, in contrast to the situation with the coaxial line, is destined to be *longer* than the free-space wavelength.

First, it can be found in handbooks on microwave technology that certain standardized rectangular waveguides are available. Assume that from considerations of power, cost and hardware parameters, we choose a guide in which the wide dimension is 7 cm. This alone determines the cut-off frequency λ_c, which is twice this value or 14 cm. (The smaller rectangular dimension only affects power-handling capability and suppression of undesired higher propagation modes. In practice, the smallest dimension would be selected consistant with power-handling capability.)

We now have sufficient information to compute the velocity factor. Because of the air dielectric, ε is 1, and the fraction under the radical sign is 10/14. Performing the indicated arithmetic operations then yields 1.43 for the velocity factor. Thus, the guide wavelength is 14.3 cm. This is tantamount to stating that the physical length of such a guide needed to transport the frequency corresponding to the magnetron's free-space wavelength of 10 cm is 14.3 cm. And, it follows that a quarter-wave section of the guide will be sized at 14.3/4 or 3.75 cm. (For practical reasons, a microwave designer is likely to make such a section several or more odd quarter-wavelengths in physical length. For example, a section 3 × 3.75 or 11.25 cm could function electrically in a similar fashion to the diminutive line of just one-quarter wavelength. The information we have derived could, of course, be used to make half-wavelength sections.

Keep in mind that the foregoing discussion assumes the use of the dominant mode. This will be the situation in most practical applications. In an experimental set-up with a variable frequency oscillator, the dominant mode would be found to be the lowest frequency that could be propagated through the waveguide. In rectangular guide, the dominant mode is defined as the TE_{10} mode. In order to avoid excessive attenuation, the wide dimension of rectangular guide is usually on the order of 0.7 λ_o.

Resonant cavities

The simplest resonant cavities are essentially entirely closed sections of coaxial cable or more often, rectangular or cylindrical waveguide. At least one dimension, often the length, must accommodate a half-wavelength in order to allow a standing-wave pattern to exist. Such a standing-wave pattern allows an appropriately placed probe or loop to sample either a very low or a very high impedance over a narrow band around the resonant frequency. The very low impedance, corresponding to series resonance, can be of the order of tens of mΩ. More useful in practice is the very high impedance which also centres around a very narrow bandwidth. This behaviour simulates parallel resonance and can manifest impedance levels of hundreds of thousand of ohms. Another way of relating these facts is to describe the cavity resonator as developing exceedingly high Q levels.

If the cavity is made from a metal which retains its physical dimensions despite temperature change and which is otherwise mechanically rugged, a frequency-governing element for oscillators is obtained which is comparable to the frequency stabilization provided at low frequencies by quartz crystals.

However, the resonant cavity does not function as simply as an LC tank circuit where one can usually feel secure that there will be just one unique and predictable resonance. A whole series of alternate parallel and series-type resonances can be expected from resonant cavities in much the same way that a length of transmission line will display resonant behaviours corresponding to odd and even quarter-wavelengths as the frequency is increased. And, complicating the matter further, there can be resonances involving different propagation modes. A cavity shape and size which will favour a desired resonant frequency can invoke a blend of art and science. The general idea is to design for operation at the lowest frequency possible for the geometry of the cavity. This is the same as designing waveguide for propagation in the dominant mode. In our favour is the fact that microwave oscillators themselves tend to favour operation at the lowest possible frequency. That is where the active device usually develops greatest gain. Also, if our luck holds out, the Q of cavity resonances usually tends to diminish at the successively higher frequencies.

Quartz crystals

Quartz is one of the crystalline substances that is known to exhibit the piezoelectric property. Such crystals undergo a change in physical dimensions when subjected to an electric field; conversely, they generate a voltage when subjected to physical deformation such as might result from application of a pressure or impact (see Fig. 1.26). Slabs or wafers cut from the body

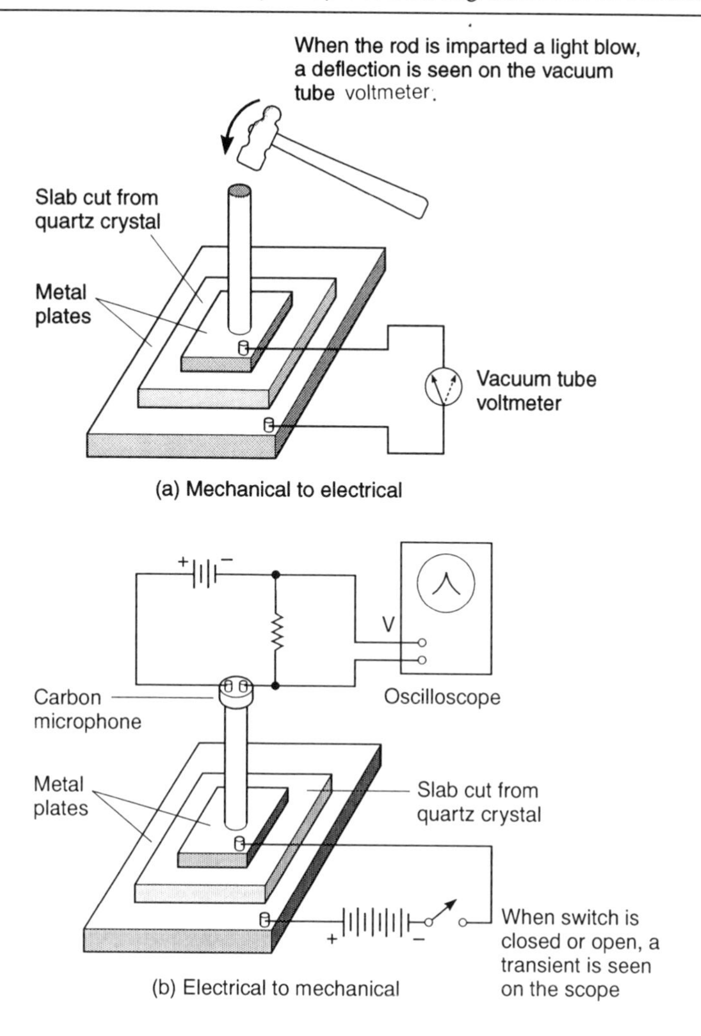

Fig. 1.26 *Experiments for demonstrating the piezoelectric effect*

of a large crystal retain the piezoelectric property. A wafer of such quartz possesses the mechanical characteristics of mass and compliance. We can therefore expect that a quartz wafer must have a natural oscillation frequency determined by its dimensions. This is, indeed, true. In the simplest and most common case, the natural oscillation frequency (mechanical resonance) is inversely proportional to the thickness of the slab, that is, the thinner the slab, the higher the frequency. Such oscillating elements are commonly called 'crystals' although they are, in reality, sections cut from a crystal. Piezoelectricity and crystallography constitute extensive studies. It

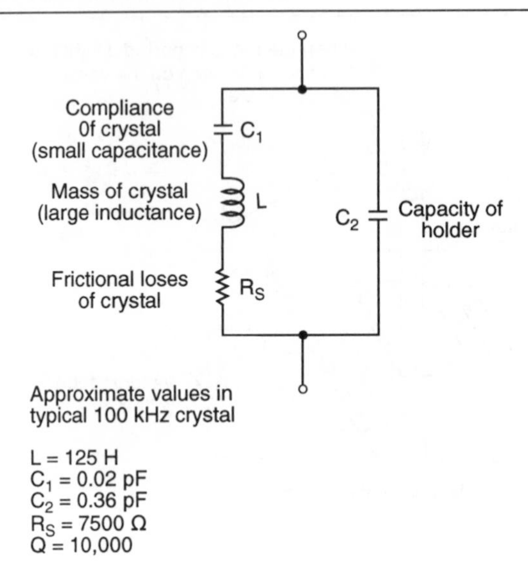

Fig. 1.27 *Equivalent circuit of a quartz crystal*

will suffice for our purpose to appreciate that the crystal can be mechanically oscillated by the application of electrical signals and, conversely, mechanical oscillation will generate electrical signals. We say that there is a *coupling* between the mechanical and electrical properties.

The two resonances in quartz crystals

The quartz-crystal oscillating element is capable of behaving either as a parallel- or a series-tuned tank circuit. The equivalent electrical circuit of the crystal oscillating element is shown in Fig. 1.27. This circuit, we see, is a combination series- and parallel-tuned network. In Fig. 1.28 the two resonances are illustrated. These are actually very close together, but it is obvious that a crystal oscillator designed to make use of the parallel resonant mode will generate a slightly higher frequency than will be obtained from a circuit in which the frequency is governed by the occurrence of series resonance in the same crystal.

The relatively small tuning effect of holder capacitance

When the piezoelectric vibrations are excited in the parallel-resonant mode of oscillation, we may think of the high inductive reactance of L as being almost, but not completely, cancelled by the high capacitive reactance of C_1. The effectively small inductive reactance then is parallel resonated by C_2. Whether operating at series or at parallel resonance, the Q of the crystal

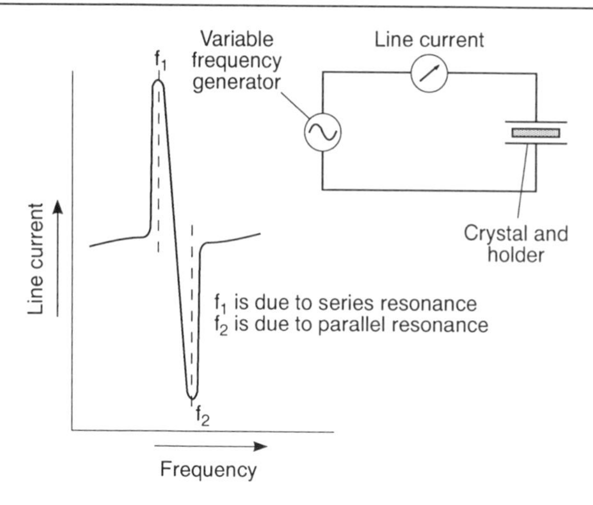

Fig. 1.28 *The two resonances, f_1 and f_2, of a quartz element*

greatly exceeds that attainable from tank circuits composed of physical inductors and capacitors. It is not possible to construct a coil with the high-Q inductance as respresented by the crystal, for either series or parallel resonance.

When the quartz crystal oscillating element is series-resonated, it appears as a very large inductance in series with a tiny capacitor and a resistance, this series network then being shunted by the relatively large holder capacitance. The holder capacitance is not part of the equivalent series-resonant circuit and cannot exert a direct tuning influence. The capacitance of the holder, and also other capacitances that might appear in this circuit position, cannot exert a pronounced tuning effect. (Series-connected capacitors produce a resultant capacitance given by the relationship, $(C_1 \times C_2)/(C_1 + C_2)$. If C_2 is, say, ten times larger than C_1, a little mathematical experimentation will show us that C_2 can be doubled with only a slight effect upon the resultant capacitance.) Although the frictional losses of the vibrating crystals are relatively high, the very high mass enables attainment of high Q.

Conditions for optimum stability

Crystal oscillating elements are capable of maintaining frequencies stable from within one part in a million to better than one part in a hundred million, depending very much upon the precautions taken to reduce or compensate the effects of temperature variations. Initially, this precaution is met by detailed attention to the angle at which the wafer is sliced from the body of the crystal. Different cuts have different coefficients of frequency change with respect to temperature. It is possible to cut the wafer in such a way as to have a positive, negative or zero temperature coefficient. A positive coefficient denotes increase in frequency as the temperature is increased. More generally, the zero temperature coefficient is considered most desirable. The second method of securing maximum frequency stability from the quartz oscillating element is to operate it at near-constant temperature inside a thermostatically controlled oven. Crystals are commonly associated with low-power oscillators wherein their salient contribution is stabilization of the generated frequency with respect to circuit conditions, temperature and mechanical shock or vibration. The crystal wafer cannot be excited too vigorously or it will shatter from excessive mechanical forces or be burned by excessive current that flows through it in the same manner as high-frequency a.c. flows 'through' any capacitor.

A closer look at crystal operating conditions

The resonance diagram (Fig. 1.29) is commonly encountered and suffices to convey a qualitative understanding of the two operating modes of oscillating crystals. However, in practical crystal oscillator circuits, parallel mode operation does not correspond to the anti-resonant frequency of the crystal in which its effective impedance would be at its maximum and would be purely resistive. Instead, the 'parallel resonant' crystal operates somewhere in its inductive reactive region. This is why parallel mode crystals are specified to work into a standardized capacitive 'load' of 32 pF. On the other hand, crystals intended to operate in the series mode of resonance can be expected to produce their stamped frequency regardless of the reactance placed across the terminals of the crystal.

It is the oscillator circuit that primarily determines whether a crystal will oscillate in its series or parallel resonant mode. However, the manufacturer can process the crystal and its mounting technique so that one or the other mode occurs with optimum results. The important thing to note is that there may be considerable departure from the stamped frequency if a crystal is operated in other than its intended mode. For some applications, this will be of no consequence; for others, the ultimate frequency, especially if derived by multiplying, will be 'out of the band'.

Δf in Fig. 1.29 has a special significance. It is the difference between the series resonant frequency and the anti-resonant frequency due to holder capacitance alone. The bandwidth of the crystal is known as f and represents the maximum frequency 'pulling' theoretically attainable by associating variable series or parallel reactance with the crystal. Other things being equal, high-Q crystals are less amenable to frequency pulling than their lower Q-counterparts. This statement may have to be somewhat tempered in practice because the lower activity of a low-Q crystal may not allow reliable oscillation over the anticipated bandwidth. Also, it must be kept in mind that lower Q means lower frequency stability.

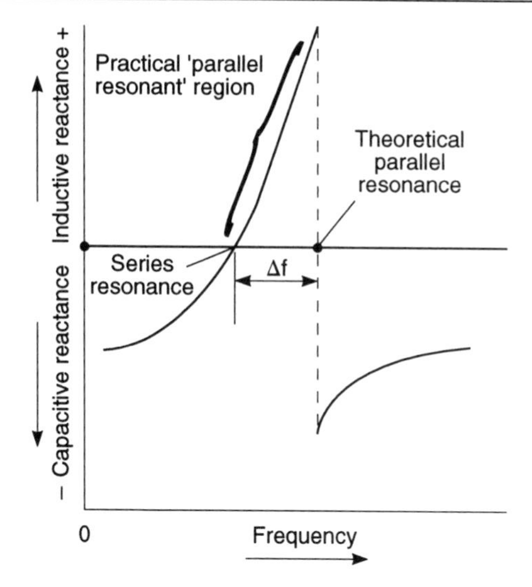

Fig. 1.29 *A closer look at the operating conditions of crystals in practical oscillators. Crystals which are spoken of as operating in the 'parallel resonance' mode actually oscillate in a frequency region in which they appear as an inductive reactance*

In order to be able to conveniently take advantage of the crystal's bandwidth in applications where it is desirable to vary the frequency, the ratio C_2/C_1 should be as low as possible. (From Fig. 1.27, C_2 is the holder capacitance and C_1 is the equivalent series capacitance of the oscillating crystal.) The best the user can do is specify a low holder capacitance and to minimize circuit and stray capacitances that essentially increase the effective holder capacitance. If these matters are not implemented, an inordinately large tuning capacitor will be needed to change the frequency across the crystal's bandwidth.

Frequency pulling in crystal oscillators

An important use of crystals is found in the VXO, the variable frequency crystal oscillator. Indeed, many 'fixed frequency' crystal oscillators are provided with a frequency trimmer. The fact that a crystal oscillator can be made tunable seems a contradiction to the widely-held notion that crystal-stabilized frequencies are 'rock-bound'. The resolution of this dilemma is in the relatively small amount the frequency of an oscillating crystal can be 'pulled'. Although small, useful results stem from this phenomenon. For example, such controllable variations of frequency are directly beneficial in amateur communications where even a slight increase or decrease in carrier frequency can lessen the interference of other stations. In receivers, the ability to fine-tune beat-frequency oscillators and 'fixed' local oscillators helps optimize reception. The fact that the frequency of a crystal can only be pulled a small percentage become less restricting when crystal frequencies are multiplied to VHF and UHF bands. This also facilitates frequency modulation inasmuch as the deviation increases along with the frequency multiplication factor; thus 500 Hz deviation at 14 MHz becomes 5 kHz deviation at 140 MHz. And frequency pulling has long been used for frequency standard and instrument time-bases where a small variable capacitor, or sometimes an inductor, has been used to set the crystal frequency.

The practical aspects of frequency pulling remain an art as well as a science. Not all crystals are equally amenable to having their 'natural' frequency varied. Much depends upon the cut of the crystal, its activity, its holder capacitance and several elusive factors. Sometimes an otherwise 'rubbery' crystal causes trouble with spurious frequencies if it is pulled too far. Obviously, the oscillator circuit and its practical implementation are very much involved. If stray capacitance or inductance is needlessly high, the pulling range tends to be narrowed. And if strong positive feedback is not available, the crystal oscillator will not start when 'detuned'.

Crystals that oscillate in their series-resonant mode can have their frequency increased by the insertion of a small series-connected variable capacitor. A typical example of this is seen in the Clapp version of the Colpitts oscillator. Parallel-resonant crystals can have their frequencies lowered by means of a small shunt-connected variable capacitor. Also, crystals oscillating in their series-resonant mode can have their frequencies *lowered* by small series-connected inductors. Generally speaking, overtone crystals are not good candidates for frequency pulling. An exception, however, is when the overtone crystal is deliberately used in its fundamental-frequency mode. In this instance, there appears to be evidence that it is capable of exceptionally-wide frequency pulling.

The crystal frequency pulling phenomenon lends itself very advantageously to varactor tuning, a technique particularly suitable for remote tuning and for frequency modulation with an audio signal. The varactor is essentially a reverse-biased pn diode which provides a useful range of capacitance in response to varying d.c. bias voltage. One of its attributes for crystal frequency-pulling is the inordinately-small minimum capacitance available.

Some practical considerations of oscillating crystals

Older crystals (those commonly found on the surplus market) were physically larger than their more modern counterparts, using natural quartz and a pressure-mounting technique in a gasket-sealed package. These were capable of safe operation at higher excitation currents than later crystals and tended to be more amenable to frequency pulling than are modern solder-seal types using synthetic quartz. However, the other side of the coin is that modern crystals are more reliable. They are less likely to be plagued by ageing effects, by decline in activity or by spurious emissions. At the same time, their temperature coefficient is either smaller or more predictable. By appropriate selection of a modern crystal and its oscillating circuit, more than enough frequency pulling is attainable for most purposes. At best, frequency pulling is a vernier effect and cannot compete with the VFO where wide-tuning ranges are needed. The frequency-pulling range for many practical applications is limited to about 1/500 of the minimal crystal oscillation frequency. Attempts at greater pulling may seriously sacrifice frequency stability.

Certain oscillator circuits, such as the Pierce oscillator and some of those used with logic ICs, will produce self-excited oscillations without the crystal being inserted in its socket. When frequency pulling is used with such circuits, attempts at excessive pulling may result in an abrupt jump to the self-excited frequency. On the other hand, circuits such as the conventional Colpitts will simply go dead if the demanded frequency pulling is too great. Even here a word of caution is in order: the Colpitts circuit is generally a ready oscillator for either series-resonant or parallel-resonant crystals. Thus, it can happen that excessive frequency pulling can cause an abrupt transition between these modes with an attendant discontinuity in the oscillation frequency. This is not likely to happen, however, unless a crystal intended for parallel resonance is operated in its series-resonant mode or vice versa.

Crystals intended for crystal filters are less active than oscillator crystals and are generally not good candidates for frequency-pulling applications. This is unfortunate inasmuch as these crystals have the otherwise desirable feature that they are remarkedly free of spurious frequency tendencies. Finally, the manufacturer's data pertaining to temperature coefficient, drift and Q pertain to 'on-frequency' operation. For demanding applications involving frequency pulling, it is probably best to consult the manufacturer who has tricks of the trade to optimize frequency-pulling ability, as well as other crystal parameters.

From a mechanical standpoint, there are several ways in which a crystal can vibrate in order to manifest itself as a piezoelectrical resonator. The flexual mode of vibration is employed for very-low-frequency crystals, enabling operation down to 200 Hz. This vibrational mode is not likely to be encountered beyond 100 kHz where one begins to find competition between the extensional and face-shear vibrational modes. From one megahertz to the vicinity of 20 to 35 MHz, the thickness-shear mode of vibration is used and the AT and BT cuts predominate. The AT cut is significant in that it generally yields the best temperature characteristics. Finally, overtone crystals also emply AT and BT cuts, and also vibrate in the thickness-shear mode. However, the overtone crystals are processed and mounted in such a way that odd overtones of the fundamental frequency are readily excited in appropriate circuits. Such crystals are practical to the ninth overtone, and may yield frequencies up to about 250 MHz.

The above information is a miniscule condensation of a voluminous subject. Fortunately, one does not have to be versed in crystallography to design and use crystal oscillators. At the outset, however, it should be known whether the oscillator circuit makes use of series or parallel resonance in the crystal. Although most crystals, regardless of cut or vibrational mode, can be operated in either the series or the parallel resonant condition, the best use of a crystal stems from operation in the resonant mode prescribed by the manufacturer. Such intended operation usually produces the highest Q, the best stability, the easiest starting, the most favourable temperature coefficient, and the closest operation to a stipulated frequency.

Some circuits obviously require a certain type of resonance. For example, the Butler and Pierce oscillators need crystals optimized for their series-resonance mode. The Miller oscillator (otherwise known as tuned-plate, tuned grid; or tuned collector, tuned base; or tuned drain, tuned gate) has to have parallel-resonant crystals. In these instances, it is customary for the manufacturer to stipulate a frequency corresponding to parallel resonance with a crystal 'load' of 32 pF.

In other circuits, the mode of resonance may not be obvious. Nor can one infer series resonance just because a small capacitor is connected in series with the crystal. For example, the ordinary Colpitts oscillator will cause its crystal to resonate in its parallel mode providing the crystral is made to develop its highest activity in this mode. If, however, the crystal has been optimized to perform in the series resonant mode, the Colpitts oscillator will become a Clapp oscillator (a Colpitts with a series-resonant tank) even though this was not the designer's intent and even though it is not apparent from the topography of the circuit.

Controlling and optimizing the temperature dependency of quartz crystals

The effective Q of an oscillating quartz crystal may be made very high by appropriate processing and manufacturing techniques and its activity can be made great enough so that reliable start-up is attainable without inordinate emphasis on the active device or on loading. But this is not enough. The temperature dependency of the crystal must be given due consideration during the processing stage and also during circuit design. For best stability, both factors are important. Crystallography is a voluminous subject and will not be dealt with in this book, but the temperature behaviour of the popular AT-cut crystal provides a good insight into the situation. This is especially true when it is recognized that other crystal cuts generally have worse temperature behaviour than the AT-cut.

Figure 1.30 shows several of the numerous cuts which may be processed from the quartz crystal. The cut greatly affects temperature coefficient, vibrational mode, Q, activity, spurious responses, harmonic behaviour and

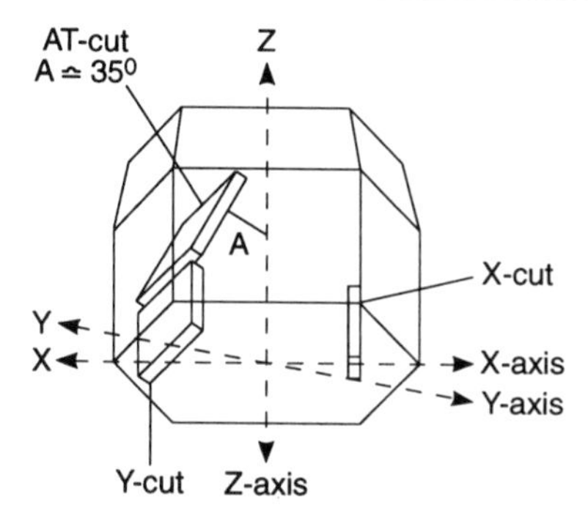

Fig. 1.30 *Temperature coefficient and other characteristics depend upon the angle of cut. The AT-cut is processed approximately at 35° with respect to the Z-axis. Minor angular deviations of this cut enables 'fine-tuning' of the temperature coefficient. The X-cut is made perpendicular to the X-axis. The Y-cut is made perpendicular to the Y-axis. Many other cuts have been used, each with its unique parameters and performance trade-offs*

frequency capability. Aside from the use of different angular cuts, oscillating crystals can assume other formats. One popular one for the approximately 32 kHz oscillator in electronic watches makes use of a quartz element in the shape of a tuning fork. The AT-cut is at a nominal 35° with respect to the *Z*-axis of the quartz crystal. However, slight angular deviations of the cut enables the manufacturer to 'fine-tune' the all-important temperature coefficient. Figure 1.31 shows how this comes about.

Not only can a substantially-zero temperature coefficient be achieved over a useful temperature range, but it is possible to design in a temperature coefficient that will then cancel the temperature coefficient of the rest of the oscillator circuit. In this way, the overall temperature coefficient of a crystal oscillator can be made very small.

Two other ways are useful in combating the temperature dependency of crystals. Circuit components, usually capacitors, are useful for this purpose because such capacitors are available with specified temperature behaviour. The idea, of course, is to cancel the temperature coefficient of the crystal, or better still, of the crystal oscillator circuit. Another long-used technique is to enclose the crystal in a temperature-stabilized oven. To be useful, the oven temperature must be above the ambient temperature. Ovens that cycle on and off have been used, but proportional types give more refined results. Sometimes it is expedient to include other oscillator components within the oven and in certain cases, the entire crystal oscillator circuit may be oven stabilized.

After any or all of the alluded safeguards against temperature dependency are incorporated, several design considerations are still needed for attainment of best frequency stability. The crystal should not be driven too hard and should be 'decoupled' as much as possible from the active device.

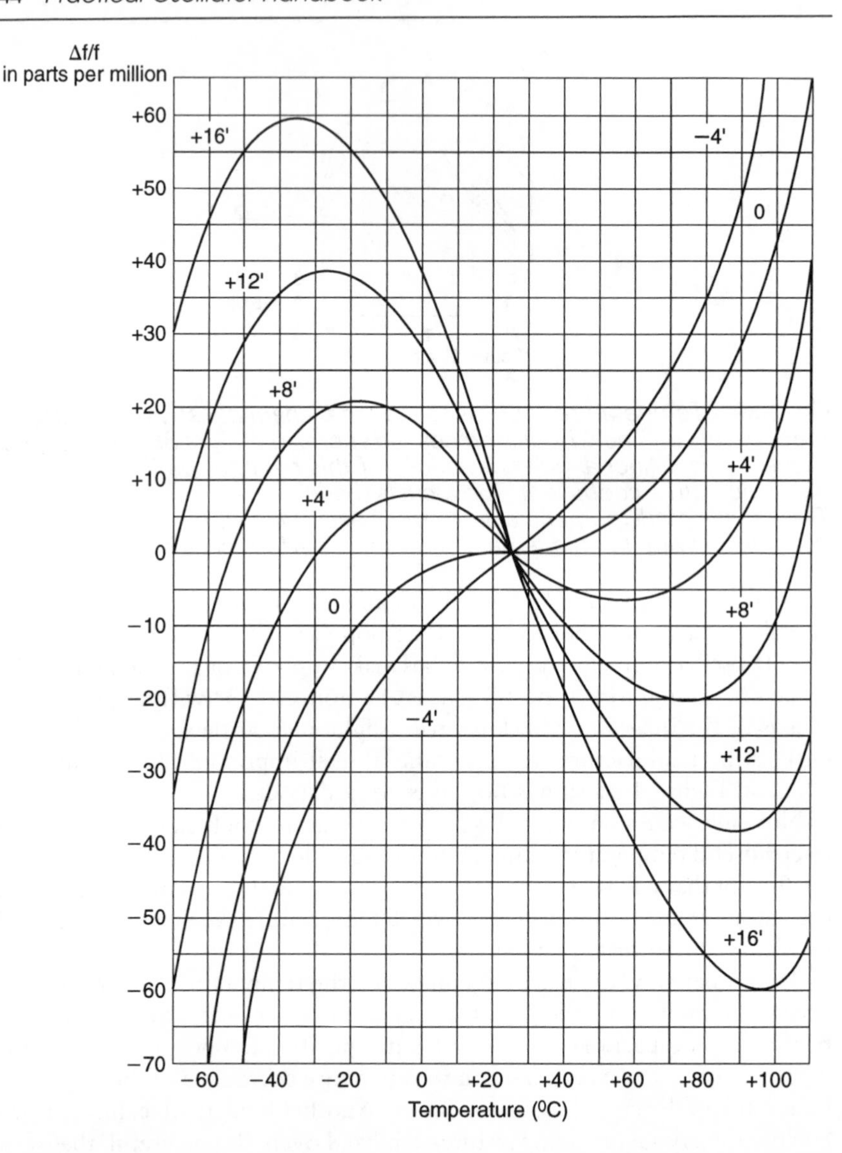

Fig. 1.31 *Temperature characteristics of AT-cut crystals versus slight changes in cut-angle. By appropriate selection of the angle of cut relative to the Z-axis, crystals optimally suited for room temperature or oven operation can be produced. Also, crystals with temperature coefficients which cancel the known temperature behaviour of the oscillator circuity can be made*

Finally, a high-isolation buffer amplifier is needed to prevent the load from affecting the oscillation frequency.

The dielectric resonator

It has long been known that the dimensions of a microwave resonant cavity could be dramatically reduced by filling it with a dielectric or insulating material. Indeed, a block or slab, or other geometric shape of such material can demonstrate resonant behaviour without being encased in metal. However, materials available until recently have been characterized by high loss, physical instability and high temperature coefficients. Previously, crystalline rutile and strontium titanate were the best dielectric materials available for microwave and millimetre-wave dielectric resonators. These materials exhibited temperature coefficients of 1000 parts per million per °C and greater. Now available are certain barium tetratitanate materials that show temperature coefficients in the vicinity of −50 at 10 GHz. The dielectric constant of such material is 38. This is an important factor inasmuch as the size of a dielectric resonator is inversely proportional to the square root of the material's dielectric constant.

The resonant frequency of a dielectric resonator is a function of its dielectric constant, its geometry and its boundary conditions. This brings us to an interesting aspect of the newer materials developed for dielectric resonators. It turns out that temperature-induced variations in dielectric constant and in physical dimensions are such that they nearly cancel. That is why very low temperature coefficients of frequency change is attainable. At the same time, dielectric losses in the new materials are so low that unloaded Qs of 6000 and better are attainable. Although the dielectric resonator cannot yet be said to compare well with quartz crystals at much lower frequencies, it must be remembered that material technology and circuit techniques are much more difficult at microwave and millimetre-wave frequencies than at the more familiar radio frequencies.

The functional diagrams (Fig. 1.32) show the use of the dielectric oscillator in microwave oscillators. The negative resistance oscillator (a) may be a Gunn or IMPATT diode, whereas the active element (b) is likely to be a gallium-arsenide field-effect transistor. In both cases, the dielectric resonator behaves as a narrowband filter that is coupled to the microwave circuitry as depicted in Fig. 1.33.

Magnetostrictive element

The piezoelectric characteristic of the quartz crystal is not the only medium whereby the mechanical and electrical properties of a vibrating element can be coupled so that mechanical movement generates an electrical signal and

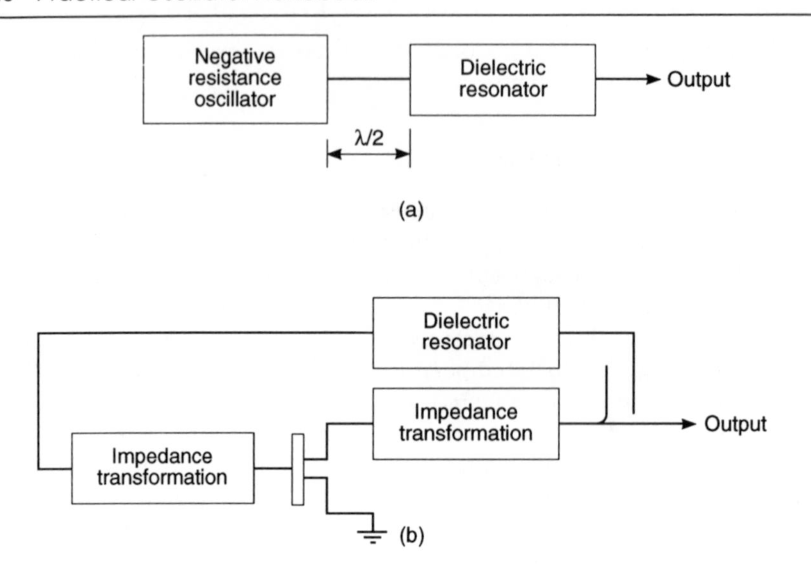

Fig. 1.32 *Deployment of the dielectric resonator in two types of microwave oscillators. a, Negative resistance oscillator. b, Feedback oscillator*

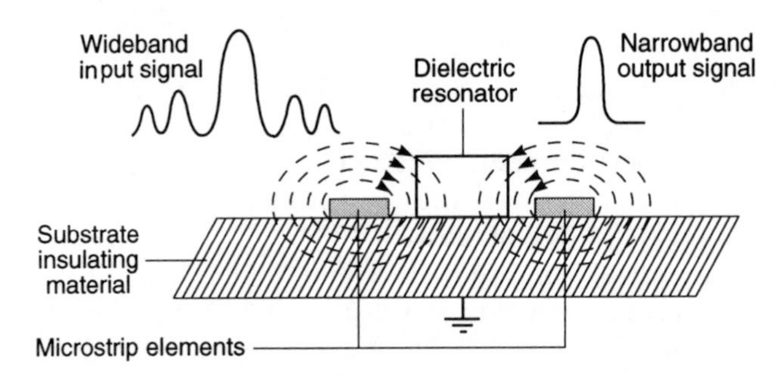

Fig. 1.33 *Coupling mechanism and filtering action of a dielectric resonator. The radio frequency magnetic fields provide energy linkage into and out of the dielectric resonator, which behaves as a selective filter*

vice versa. A somewhat analogous situation is found in the magnetostrictive effect in certain magnetic materials, especially in nickel alloys. Magnetostriction is the change in length produced in rods of such materials in response to a magnetic field, and the converse change in magnetization that results when such rods are subjected to mechanical tension or compression. The word is misleading in that it suggests a shortening of length only. Actually, both shortening and lengthening exist in different alloys, and in a few instances, in the same alloy (see Fig. 1.34). With most magnetostrictive alloys, the shortening (or lengthening) will occur for each half of the a.c. cycle. This means that the rod undergoes two complete cycles of mechanical vibration

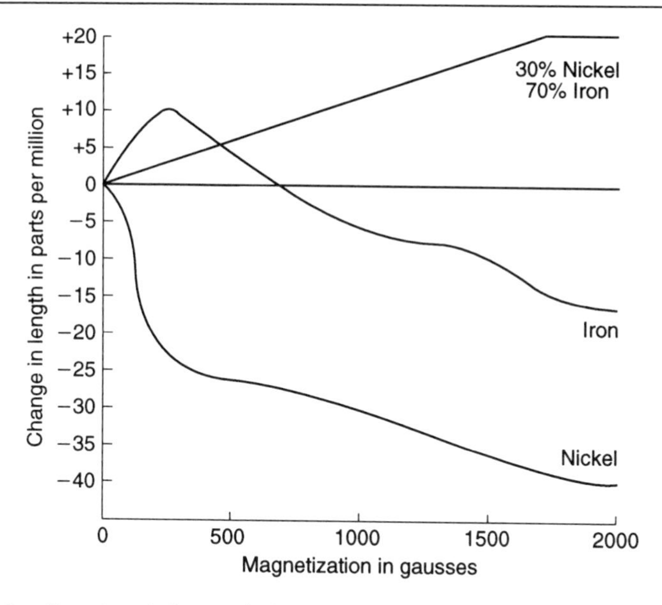

Fig. 1.34 *Fractional change in length caused by magnetostriction effect in several materials*

for each cycle of the exciting force. (This did not happen with the piezoelectric crystal because the crystal becomes thicker for one half-cycle of applied potential, then thinner for the second half-cycle. In other words, the vibrations of the crystal 'follow' the electrical oscillations.) It will be shown that the magnetostrictive element is suggestive of a transformer with very loose coupling at off-resonant frequencies, but with tight coupling at resonance.

Need for bias

The frequency-doubling property of the vibrating magnetostrictive rod is generally not desirable. In order to make the rod maintain pace with the applied magnetic oscillation, it is necessary to superimpose a constant magnetic force. This is readily obtained from the effect of a direct current that is either applied to a separate bias winding or allowed to circulate in a signal winding. This magnetic bias prevents reversal of the sense of magnetization, thereby causing the rod to vibrate at the same frequency as the signal (see Fig. 1.35). An analogous situation is found in the ear-phone, which must be provided with a small permanent magnet so that the magnetic force acting on the diaphragm does not reverse its sense. Without such provision, severe distortion would occur because the diaphragm would tend to vibrate at a frequency twice that of the electromagnetic signal representing the speech or music.

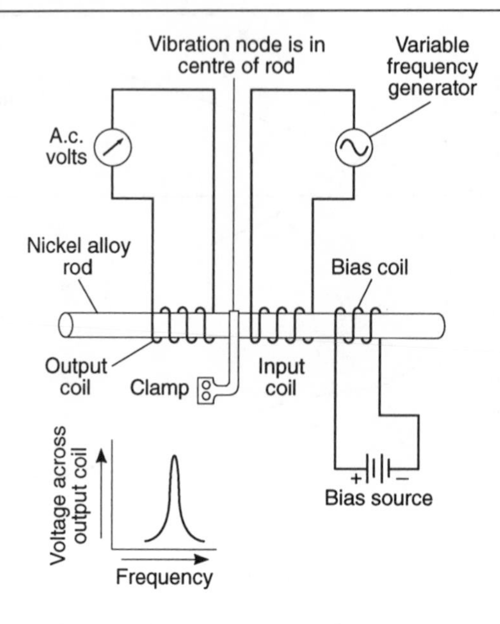

Fig. 1.35 *The magnetostrictive oscillating element*

Frequency/length relationships in magnetostrictive element

A nickel alloy rod about four inches in length vibrates at a frequency of 25 kHz. The frequency of resonant vibration is inversely proportional to length so that practical limitations are encountered at both low and high frequencies. We see that a 100 kHz magnetostrictive element is on the order of an inch in length, whereas a 20-inch rod is required for 5-kHz oscillation.

The magnetostriction oscillator

When connected in conjunction with a suitable oscillation-provoking device, such as a vacuum tube, an a.c. signal applied to an exciting coil subjects the biased rod to a magnetic force of cyclically varying strength. This causes the rod to change its length accordingly. The changes in rod length are accompanied by changes in its magnetization that in turn induce a signal in another coil. The induced signal is of maximum amplitude at the frequency corresponding to vibrational resonance of the rod, for it is then that the changes in length and magnetization are greatest. The signal undergoes amplification in the vacuum tube, or other device, then is reapplied to the exciting coil of the magnetostrictive oscillating element. In this way, oscillation is maintained at the mechanical frequency of resonance in the rod. When the magnetostriction element is associated with an amplifier, phasing of the windings is made opposite to that necessary for conventional 'tickler' or Hartley feedback oscillators. This is true because induced voltage from magnetostriction is, at any instant, of opposite polarity to voltage generated by ordinary transformer action. This is fortunate, for otherwise it would be very difficult to prevent oscillation at an undesired frequency due to electromagnetic feedback.

The tuning fork

Besides the quartz crystal and the magnetostrictive rod, a third electromechanical oscillating element is of considerable importance. This is the electrically driven tuning fork. A steel bar bent into a U shape can be excited into vibration by means of a solenoid placed near one of the fork ends. As we might expect, the maximum intensity of the vibration occurs when the frequency of the solenoid current is the natural resonance frequency of the fork. The region near the U junction is a nodal region where vibrational amplitude is at a minimum. On the other hand, the ends of the fork prongs execute transverse vibrations of relatively great amplitude. Suppose we place a second solenoid at a corresponding position near the opposite prong. If, then, we monitored the induced voltage in this winding, we would find that it peaked up sharply at the resonant frequency of the fork (see Fig. 1.36). The flux linking the pick-up winding undergoes its most intense variation at resonance due to the mechanical motion of the fork. This, in turn, produces the greatest change in flux in the pick-up winding thereby causing the observed rise in a.c. voltage developed in the pick-up winding. When the fork is connected to an amplifier to form an oscillating circuit, the pick-up winding delivers a signal to the input of the amplifier. The amplified signal appearing at the output of the amplifier energizes the exciting solenoid. The resultant magnetic impetus then serves to regenerate this oscillation cycle and sustained oscillation ensues.

Applications

Over the frequency range of about 100 Hz to several tens of kilohertz, such a fork behaves as a high-Q energy storage system capable of providing excellent frequency stability. Other types of tuning forks have been used. In one type, the fork carries contacts and the device becomes essentially a buzzer. In another version, the pick-up signal is obtained through the use of a carbon microphone.

RC networks as oscillating elements

Numbered among oscillating elements are certain combinations of resistance and capacity. Such networks are not 'natural' supporters of oscilla-

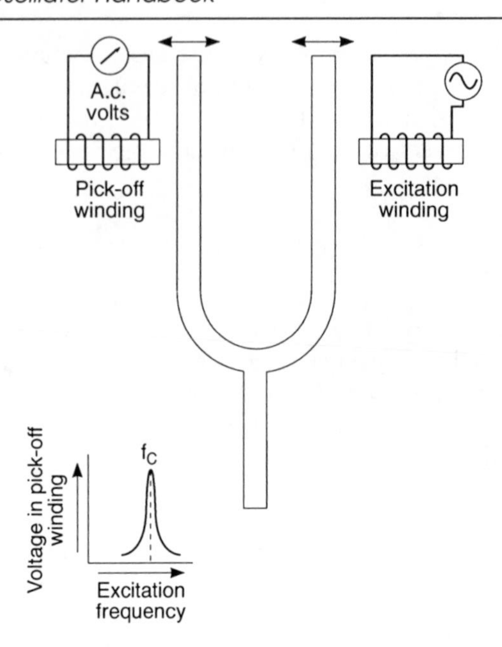

Fig. 1.36 *The tuning fork*

tions; rather, they become oscillating elements in special amplifier circuits. Even in those RC networks in which sharp frequency selectivity exists, the tendency to 'ring' that characterizes LC tank circuits is entirely absent. (However, ringing can be demonstrated with the network connected to its feedback amplifier.) The RC network, no matter what its form, functions essentially as a phase-shifting circuit. Generally, the oscillator circuit operates at the frequency at which the RC network produces a 180° phase shift or zero, whichever corresponds to positive feedback. Other frequencies produce phase shifts that correspond to negative feedback. As a consequence there is high discrimination against such frequencies. This permits generation of a very nearly pure sine wave. To accomplish the same thing with an LC tank, the Q would have to be quite high.

One of the useful properties of the RC networks is that frequency is inversely proportional to capacitance, rather than to the square root of capacitance as in an LC tank circuit. This means that a given change in capacitance produces a much greater frequency change than in an LC tank circuit. If we increase capacitance nine times in the LC tank, we decrease the resonant frequency by one-third. The same increase in capacitance in RC networks decreases frequency by one ninth. It is, however, often necessary to simultaneously tune two or three elements in the RC network. An additional disadvantage is the fact that the output power of the RC oscillator is necessarily low by virtue of the dissipation in the resistive elements.

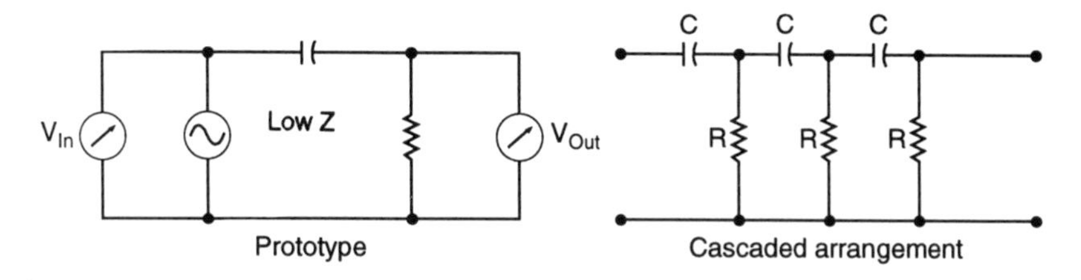

Fig. 1.37 *RC phase-shift network*

Examples of RC networks suitable as oscillating elements

One of the simplest means of obtaining phase inversion (180° phase shift) is to cascade three similar RC high-pass filter circuits. The resulting phase-shift network is shown in Fig. 1.37. The simple prototype RC high-pass network is limited in its phase-shifting capability. Extended phase shifting is obtained by cascading two or more of the prototype circuits. When such a network is connected between input and output of a single-tube amplifier, oscillation occurs at the frequency at which the network compensates the natural 180° phase shift in the amplifier. For three cascaded RC pairs as shown, a 180° phase shift occurs at a frequency approximately equal to:

$$f = \frac{1 \times 10^6}{2\pi\sqrt{6}\ RC}$$

where:

f is the frequency in hertz,
R is the resistance in ohms,
C is the capacitance in microfarads.

A similar phase inversion is produced at the 'resonant' frequency of the parallel-T bridge illustrated in Fig. 1.38. For this bridge, resonant frequency occurs at:

$$f_0 = \frac{1}{2\pi\ (C_1)\ (R_2)}$$

$$= \frac{1}{2\pi\ (C_2)\ (R_1)}$$

where:

f_0 is the frequency in hertz,
R_1 and R_2 are in ohms,
C_1 and C_2 are in farads,
R_2 equals $2R_1$,
C_2 equals $2C_1$.

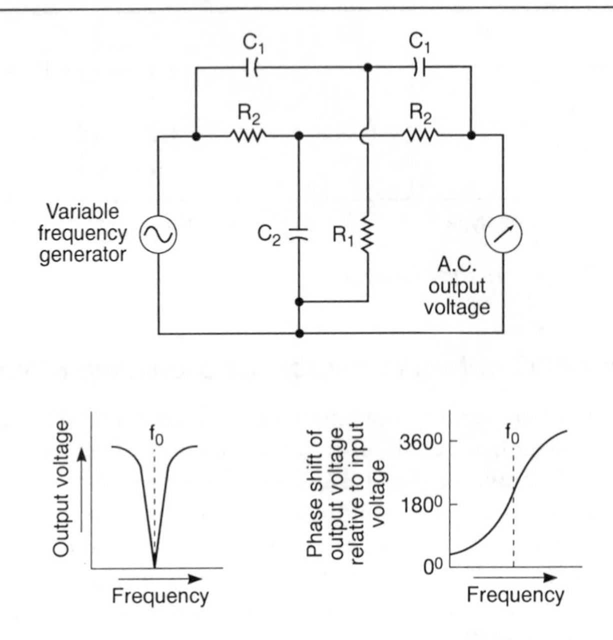

Fig. 1.38 *The parallel-T network*

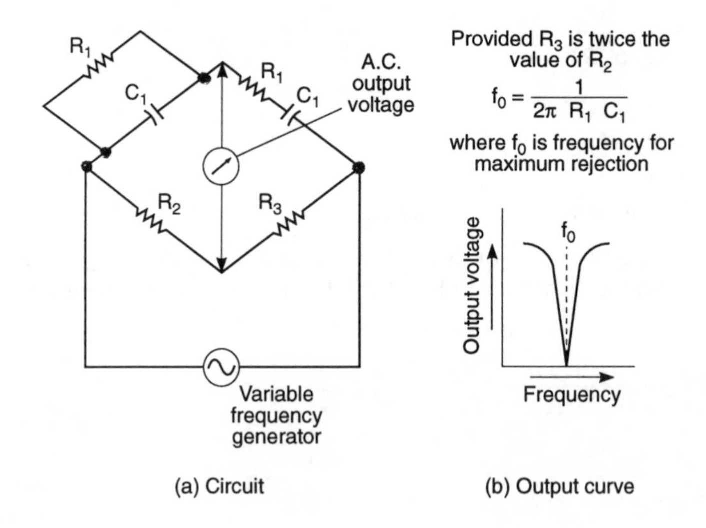

Fig. 1.39 *The Wien bridge*

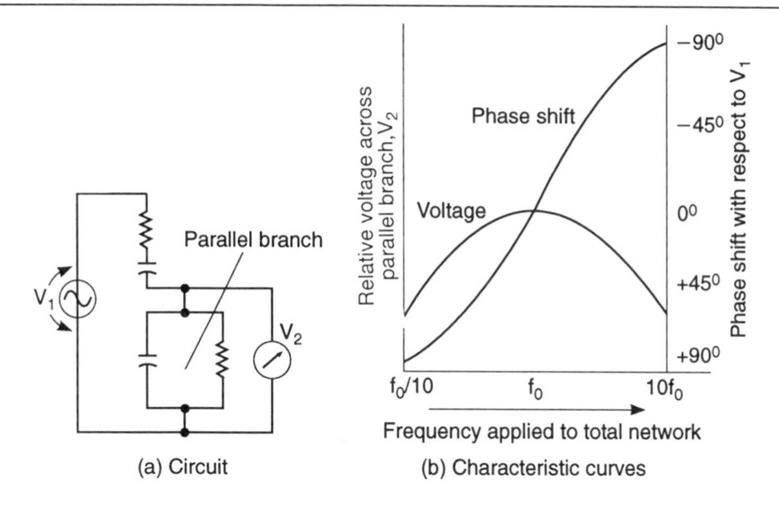

Fig. 1.40 *Characteristics of reactive arms of Wien bridge*

In this case, the phase inversion is accompanied by maximum attenuation of the signal traversing the network. This is not the desirable response characteristic because it tends to defeat our purpose if we are to invert the output signal of an amplifier, but at the same time drastically reduce its amplitude. However, it is comparatively easy to compensate for the high attenuation by providing sufficient voltage amplification in the amplifier associated with the network. When this is done, oscillation takes place at the frequency 'null' of the network. Another bridge with phase and amplitude response similar to that of the parallel-T network is the Wien bridge depicted in Fig. 1.39.

The Wien bridge is not always readily recognizable as such in schematics. This is because the draughtsman looks at the world differently from the electronics practitioner. Thus, it often happens that the arms of the bridge, instead of being lumped together in a neat diamond configuration, are more or less randomly scattered about. (Incidentally the Q of this RC network is one third.) In any event, the characteristics of the bridge are shown in Fig. 1.40. See the section on oscillator theory for more details. A modified type of this oscillator is also described in Chapter 5.

LC networks as phase shifters

Of utmost importance are the phase properties of certain LC networks. These LC networks produce phase inversion at a discrete frequency, thereby enabling a portion of the output signal of an oscillator tube to be returned to

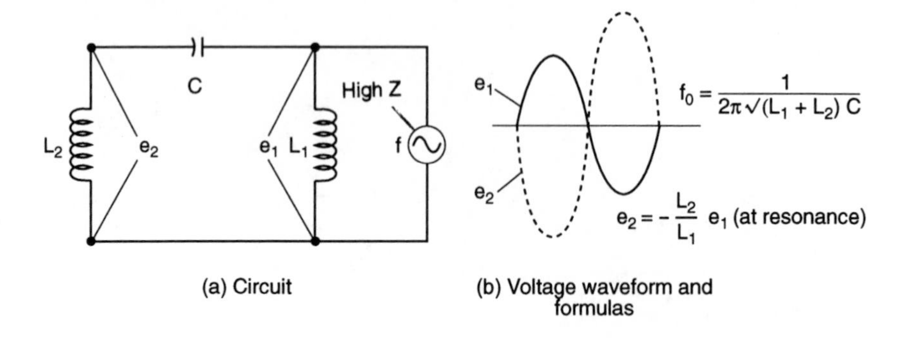

Fig. 1.41 *An LC network that produces phase inversion (Hartley type)*

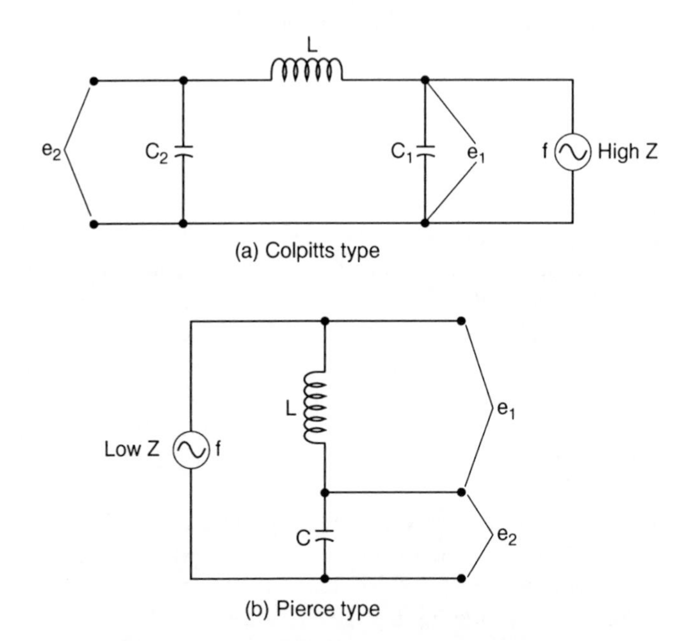

Fig. 1.42 *Additional LC networks that produce phase inversion*

the input circuit in proper phase relationship to reinforce the signal already there. The IC networks illustrated in Figs 1.41 and 1.42 are basic to the operation of the majority of LC feedback oscillators. However, they are not always immediately recognizable as being equivalent to these simple circuits. For example, what we have shown as an inductor may actually be a parallel-tuned tank operated slightly below its resonant frequency, under which condition such a tank appears inductive. Conversely, the inductor may be a series-tuned tank operated slightly above its resonant frequency.

Instead of physical inductors or inductor–capacitor combinations, an actual oscillator may employ a piezoelectric crystal. In Chapter 4 on practical oscillators, we will find these LC networks directly relevant to the basic operating principles of important oscillator circuits. For instance, the network of Fig. 1.41 underlies the theory of oscillation in the Hartley and also in the tuned-plate tuned-grid oscillator. Similarly, the principle of the well-known Colpitts oscillator depends upon the phase-inverting property of the LC network shown in Fig. 1.42a.

2 Active devices used in oscillators

An active or oscillation-provoking device is as essential to the oscillator as are frequency-determining elements. Such devices remove the effect of dissipative losses that are always present in frequency-determining elements at ordinary temperatures. This may appear to be a strange way of approaching the subject, but it is a demonstrable fact that an LC resonant 'tank' can be set into sustained oscillation at near absolute zero temperature. It is, indeed, instructive that no amplifier, feedback path, switching device, negative-resistance device or other driving means is required. At the low cryogenic temperature level, all one need do is momentarily induce the oscillation from an external source; thereafter, the LC tank continues *of its own accord* to oscillate for a very long time. (In principle, it would not oscillate forever at absolute zero because the energy would ultimately be dissipated via radiation.)

Of course, one can also argue that such sustained LC oscillation must eventually die down because some losses can be expected even at zero temperature. One such loss is the production of eddy currents in conductors in the nearby environment; another loss must be attributable to radiation. However, we need not split hairs to convey the general idea that we can bring about oscillation via some means of overcoming the inherent dissipative losses.

The three common ways of causing oscillation in ordinary practice are by associating the resonant circuit with a power amplifier, with a switch or with a negative resistance. Although these three 'provokers' appear quite different from one another, from the 'viewpoint' of the resonant tank they are just three manifestations of a common theme, i.e., all three methods essentially supply energy that can be used to overcome the dissipative losses. Thus, with appropriate circuit modifications, many active devices, such as transistors and op-amps, can be made to behave as a power amplifier, a switch or a negative-resistance, and can provoke oscillation in a given LC tank in any of these three modes. The three modes are, as one would expect, mathematically equivalent, but it is not easy to show this for practical purposes.

Practical oscillator applications are best served by discussing the salient characteristics of the most frequently encountered active devices. That is what we shall do in this chapter. Although the vacuum tube was once the dominant device, we have progressed into an era in which a number of solid-state devices offer advantages of cost, reliability, efficiency, longevity and physical compactness. Tubes continue to merit consideration in certain applications, however. Thus, some form of electron tube is often found in microwave ovens, radar, lasers and in induction-heating.

The bipolar junction transistor

The bipolar transistor is probably the simplest and the most versatile active device for oscillators. It is conveniently thought of as a current-activated three-electrode amplifier. Inasmuch as its control electrode, the base, consumes current even at direct-current operation, its input impedance is necessarily lower than voltage-actuated devices, such as the erstwhile electron tube or the MOSFET. This low impedance can seriously degrade the Q of an LC resonant circuit, so design procedure must deal with this. If the LC tank is too drastically loaded down by this low input impedance, the oscillator will exhibit problems in stability, power transference to the load and in self-starting. One commonly alleviates this effect by choosing a high-grain transistor and 'tapping down' the connection to the inductor and by isolating the LC circuit as much as possible from the 'current-hungry' base of the transitor. This can be done with a small coupling capacitor, selected so that reliable oscillation just takes place.

Most bipolar transistors are now silicon types, but some germanium transistors are still made and can be advantageously used in specialized applications. The threshold voltage for silicon transistors is about 0.6 V, but is only approximately 0.3 V for the germanium types. This defines the base-emitter voltage wherein collector current conduction just commences. Prior to attaining the required bias voltage, the input impedance to the transitor is exceedingly high, the transistor is effectively 'dead' as an active device. Once the bias voltage reaches the threshold level, input current is consumed, the input impedance becomes relatively low and collector current flows in the output of the device.

Most circuits use the common-emitter configuration, but the common-base circuit is actually more natural for the transistor. Its main shortcoming for oscillators is that the input impedance is much lower than for the common-emitter configuration. With modern high current-gain transistors, this low input impedance ordinarily doesn't prove troublesome until one deals with UHF and microwave oscillators. At these very high frequen-

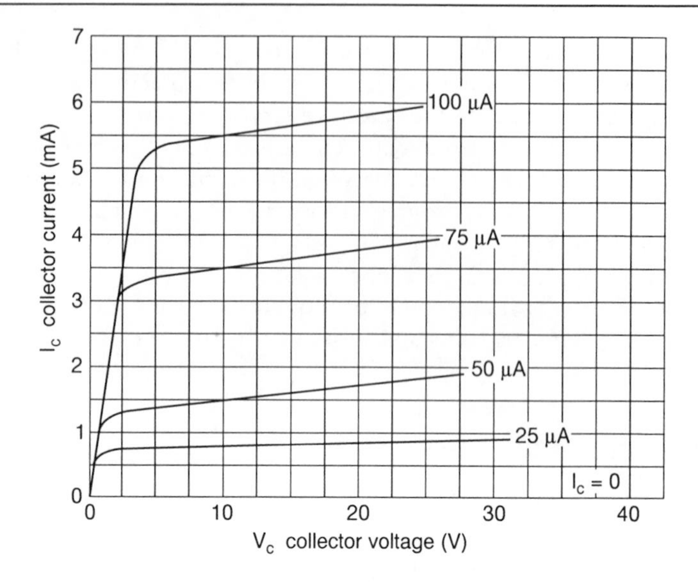

Fig. 2.1 *Common-emitter output characteristics of a typical npn transistor. This is probably the most often used format in oscillator circuits. It tends to provide a reasonable balance of such features as power gain, input impedance, temperature stability and d.c. feed convenience. A small unbypassed resistance connected in the emitter lead will often improve the waveshape of the oscillation*

cies, however, input impedance is just one of many non-ideal characteristics that must be traded off, such as lead inductance, low power gain, package capacitances, thermal limitations and device fragility. Here, the common-base circuit is likely to merit design consideration because it is much less vulnerable to 'parasitic' oscillation at unintended low frequencies than the common emitter configuration. Common-emitter and common-base output characteristics for a typical npn small-signal bipolar transistor are respectively shown in Figs 2.1 and 2.2.

Transistor polarity and Darlington pairs

Both bipolar and MOS devices are available in the two 'genders'. Thus, npn bipolar transistors operate with their collectors polarized positive with respect to their emitters or bases. In the npn transistor, 'on' bias is established by making the base *positive* with respect to the emitter. The converse situation pertains to pnp bipolar transistors wherein the collector is operated with negative polarity with respect to emitter or base. 'On' bias with these devices is brought about by making their base *negative* with respect to the emitter. Otherwise, very nearly identical operation can be obtained from

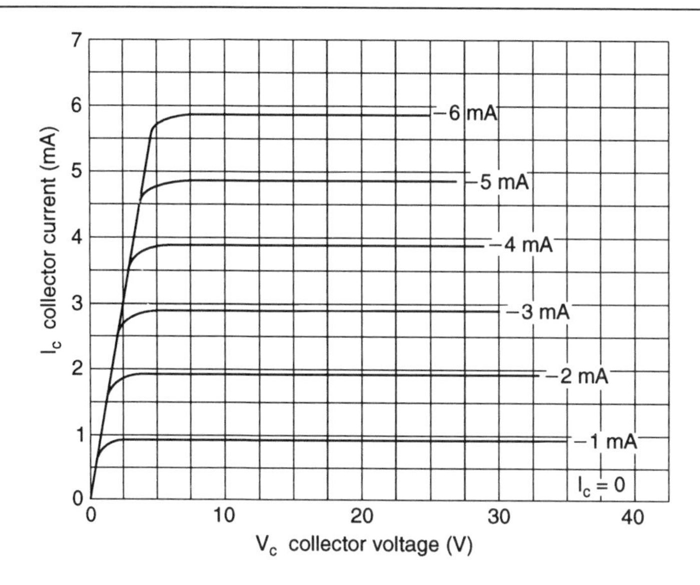

Fig. 2.2 *Common-base curves of the same transitor shown in Fig. 2.1. In this family of curves involving different emitter currents, note the negative polarity indications. This is OK for such an npn device because the base remains* positive *with respect to the emitter just as in the common-emitter connection. This graph depicts the nearly ideal behaviour of the common-base arrangement. Observe, however, that we are dealing with milliampere inputs contrasted to microampere inputs for the common-emitter circuit*

npn and pnp oscillators—it's largely a matter of serving the convenience of the d.c. power source at hand. However, there are probably a lot more npn types readily available. Not all npn types are made with a pnp matching type. Note the common-emitter output characteristics of the pnp transistor shown in Fig. 2.3.

For many practical uses, manufacturers provide npn and pnp transistors which are very similar to one another. However, a study of semiconductor physics reveals that it is inherently difficult to make true 'mirror images'. This stems from the fact that electrons and holes, which have their roles transposed in the two types, have different mobilities. Despite this, there are interesting applications of complementary symmetry making use of an npn and a pnp transistor sharing operation as the active devices. This usually results in an easily implemented push–pull circuit without need for a centre-tapped tank.

Another situation in which a *pair* of bipolar transistors comprise the active device is the use of one of the four Darlington configurations shown in Fig. 2.4. The Darlington connections feature high input-impedance, and very high-current and power gain. Oscillators configured around such Darling-

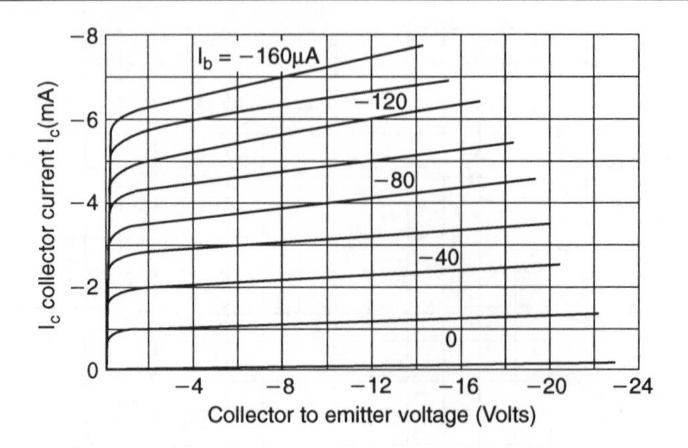

Fig. 2.3 *Typical common-emitter output characteristics of pnp transistors. The curves can be seen to be 'mirror-image' versions of those representing npn transistors. Thus, all of the involved d.c. polarities depicted above are negative rather than positive. Otherwise, npn and pnp types yield about the same performance in most oscillators. However, more npn than pnp types are available for VHF, UHF and microwave applications. (Hole mobility in the pnp device is slower than electron mobility in npn versions.)*

ton pairs seem good candidates where wave purity is important. (The output device does not saturate.) Sometimes the same simple symbol used to depict the single-device bipolar transistor is used for the Darlington. This is probably because such Darlington types are available in one package so that they physically resemble single-device transistors. Most of these Darlingtons are made for handling higher power levels, in the tens and hundreds of watts. They can be simulated for small-signal oscillators by combining appropriate discrete transistors.

Whether one thinks of the bipolar transistor as an amplifying device with high current gain, high power gain or high transconductance, it is a very willing oscillator. Good performance is attainable from very low frequencies through the microwave portion of the radio frequency spectrum if one pays heed to manufacturer's specifications. Failure modes such as thermal runaway and secondary breakdown can be reliably avoided via operation well within the boundries of maximum ratings. The relative fragility of the higher frequency types should be respected and means should be incorporated to suppress oscillation at anything other than the intended frequency.

MOSFET transistors

The n-channel MOSFET corresponds to the npn bipolar transistor in that it works from a positive d.c. power supply. That is, the drain is impressed with

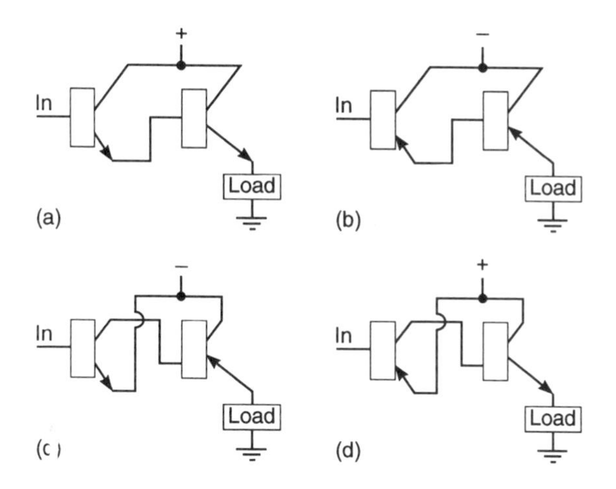

Fig. 2.4 *The four Darlington circuits using bipolar transistors. The output device may or may not be a power transistor. Because of the high input impedance and high current gain of these pairs, they are particularly suitable for RC oscillators such as the phase shift, the parallel T and the Wien. (a) npn pair, (b) pnp pair, (c) npn input, pnp output, (d) pnp input, npn output.*

positive polarity with respect to the source and conduction is brought about by applying positive gate voltage with respect to the source. It will be noted that conduction does not gradually commence from zero gate voltage, rather, there is a threshold gate voltage below which the device is completely turned off. This is suggestive of the threshold base-emitter voltage in bipolar transistors. However in the MOSFET, the threshold voltage can be widely varied by the manufacturer. Particularly useful in certain applications are MOSFETs with low threshold voltages so that they can be controlled by logic circuits. The p-channel MOSFET is intended as the mirror image of the n-channel types. The d.c. operating polarities cited for the n-channel types are reversed. That is, the drain is negative with respect to the source, and a negatively polarized gate voltage is needed to bias the device to its conductive state. Typical n-channel curves are shown in Fig. 2.5.

Power MOSFETs may have power dissipation ratings of many tens or hundreds of watts. However, experimenters have found that these devices are also ideal for low-level oscillators. The same is not ordinarily true with regard to bipolar power transistors—leakage currents, while negligible at high-power operation, tend to interfere with successful operation at low power levels. Use small-signal bipolar transistors for low-power oscillators.

MOSFET oscillator circuits do not exhibit the tendency to engage in low-frequency spurious modes as do bipolar oscillator circuits. This is

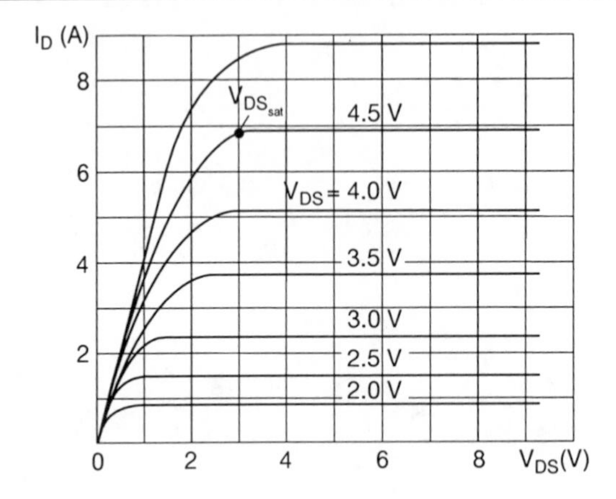

Fig. 2.5 *Output characteristics of a typical n-channel power MOSFET. Because positive gate-to-source bias is required for active operation, such devices are said to operate in the* enhancement *mode. Note the two operating regions: the sloping portion of the curves represents the* linear *region where on-resistance is approximately constant. (This is also known as the ohmic region.) The horizontal portion of the curves represents the* saturation *region where drain current is nearly constant. A point such as $V_{DS_{sat}}$ which divides these two operating-regions is established as:*

$$V_{DS_{sat}} = V_{GS} - V_T$$

where V_T is the threshold voltage given in the manufacturer's specifications. (V_T above is 1.5 V)

because MOSFET power gain is nearly uniform over extremely wide frequency ranges. However, MOSFET oscillators often simulate electron tubes in their willingness to generate very high-frequency parasitic oscillations. To guard against this undesirable performance, a non-inductive resistor of ten to a hundred ohms should be inserted directly at the gate terminal. Instead of the resistor, or in combination with it, a 'lossy' ferrite bead is often found useful in suppression of such parasitics. Although simultaneous parasitic oscillation may be benign, it can also be the agent of mysterious malperformances of the oscillator, as well as damage to the MOSFET.

Small signal MOS devices are available in dual-gate types in which control can be independently (over a large range) extended by the individual gates. This is very useful for various mixing and modulating techniques. Also, isolation is readily provided between biasing and signal circuitry in these devices. The experimenter is reminded of multi-grid electron tubes. One often finds oscillator circuits in the technical literature with the two gates of such devices tied together.

Operating modes of MOSFETs and JFETs

MOS devices are considered to be voltage actuated. In this respect, they simulate the ordinary operation of the electron tube. Their exceedingly high input impedance makes them attractive for oscillators because negligible loading is imposed on the resonant tank circuit. However, this statement, while essentially true at low and moderate frequencies, must be further qualified at VHF and higher frequencies. For one thing, the input to these devices looks like a near-perfect capacitor at low frequencies. At very high frequencies, the current consumption of this input capacitance causes losses in the input circuit. Also, the capacitance itself is no longer as 'pure' as it appears to the lower frequencies. Although still a willing oscillator as we approach the microwave region, the MOS device no longer operates as if it develops near infinite current gain as at low and moderate frequencies. Indeed, it begins to simulate the 'current-hungry' operation of bipolar transistors.

Most MOS devices are enhancement-mode types. That is, they are normally turned off when their gates are not provided with forward conduction bias. When so provided, they 'become alive' in a similar manner to that of bipolar transistors. Unlike bipolar transistors, they consume no input at d.c. Depletion-mode types are also available; these are normally conductive and the conduction can be both increased and decreased by varying the bias or by the impressed signal voltage. This is reminiscent of most of the electron tubes used in radio circuits.

The control electrode of MOS devices can be a reverse-biased diode such as we have in the JFET, or the actual conductive plate of a capacitor such as is used in both small-signal and power MOSFETS. In both cases, the input impedance is essentially capacitive. In both cases, the conductance of a 'channel' is modulated by the electric field caused by the input voltage. The output characteristics of a typical JFET are shown in Fig. 2.6.

Generally speaking, MOS devices tend to be immune to thermal runaway and to failure from secondary breakdown. This does not mean that abusive overloading will not damage or destroy them. All electrical devices can be 'burned out' from excessive temperature rise, current, or voltage. Take particular care in handling MOS devices; their thin silicon dioxide dielectrics are easily punctured by gate voltages ranging from several-tens to several-hundreds of volts. Static electricity accumulated by the body, or generated by frictional contact between various insulating materials can easily manifest itself in *thousands* of volts. That is why these devices are packed with their leads inserted in conductive foam. During experimentation and construction, grounding techniques should be used to ensure that no difference of potential can exist between the MOS device and one's body or tools. Once protected against such gate damage, MOS devices prove to be electrically rugged when properly operated in oscillator circuits.

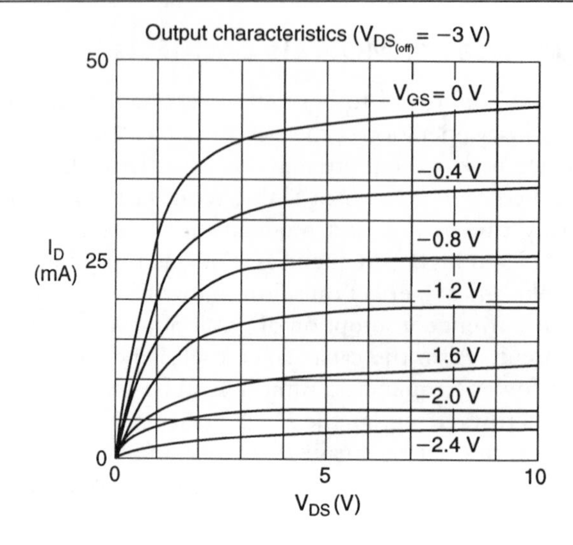

Fig. 2.6 *Output characteristics of a junction field-effect transistor (JFET). These popular small-signal devices, unlike most other solid-state amplifiers, operate in the* depletion *mode. This infers that there is strong drain current when the gate-source voltage is zero, i.e., the JFET is normally* on. *This JFET is an n-channel type inasmuch as the drain is made positive with respect to the source. All of the indicated polarities would be reversed for a p-channel JFET*

The voltage-follower format of active devices

The common-collector circuit or voltage follower, otherwise known as the emitter, source or cathode follower, is also useful in oscillators. Inasmuch as these configurations exhibit less than unity voltage gain in practical applications, a knee-jerk reaction might be that there could be no oscillation. However, when we have an LC resonant tank and a feedback path, oscillation is always possible notwithstanding less than unity voltage or current gain of an associated active device. This follows from the criterion of oscillation with regard to *power gain*, which must equal or exceed unity. It so happens that the common collector circuit can provide respectable power gain. It does this by developing close-to-unity voltage gain, but very high current gain.

One should contemplate how this situation differs from that of a *transformer.* The transformer can produce either voltage or current gain, but it never can develop power gain. Indeed, practical transformers always show less than unity power gain. There is no way that ordinary linear transformers can be associated with an LC circuit to provoke oscillation. Even the saturable-core transformer, which somewhat resembles an active device, must be associated with switching elements in order to participate in the oscillatory process.

In power transistors, the collector is often internally connected to the metal package. This ties in nicely with certain oscillator circuits where the collector is at audio or radio frequency ground potential. This enables the transistor to be mounted on a heat sink, or on the chassis with no need for insulating washers or gaskets. Such a situation allows more efficient heat removal and contributes to overall electrical reliability. Also, the common-collector arrangement often enables one to dispense with the radio frequency choke needed for shunt-fed 'hot' collector circuits.

Of the three configurations, the common-collector circuit displays the highest input impedance and therefore tends to exhibit a more benign influence on an associated LC tank. It is well to keep in mind, however, that at VHF and higher frequencies, the differences in impedance levels of the three configurations tend to be less pronounced. At microwaves, experimentation is sometimes needed to determine the optimum arrangement.

Bias considerations in active devices

In the technical literature, one sometimes finds allusions to solid-state oscillators operating in Class C. Accuracy would be better served if the statements were modified to describe the operation as 'somewhat like Class C'. In the electron-tube oscillators, Class C operation was accompanied by rectified grid current which could be indicated on a d.c. meter. Indeed, the oscillator could be adjusted for optimum performance by maximizing this grid current. One does *not* observe the counterpart of this situation in solid-state oscillators.

Consider MOSFETs, which in many ways perform similarly to tubes. Practically speaking, the gate-source input circuit of these devices behaves as a capacitor. Therefore there can be no rectified oscillation current to monitor at the gate. The JFET is a depletion device like most tubes and one can suspect the possibility of d.c. gate current as consequence of oscillation. Here, too, is an input diode which could be impressed with forward-conduction voltage (like a grid) and thereby give rise to a d.c. gate current. It turns out, however, that forward biasing the gate actually *inhibits* the operation of the device and can even be damaging. Although rectification is possible at the input of both the tube and the JFET, the results are quite different.

Bipolar transistors operate with d.c. current consumed by their base-emitter input sections. This current is *not* analogous to the d.c. current in the grid-cathode input of an electron tube. In the bipolar transistor, the d.c. input current is a measure of the forward bias needed to inject charge carriers into the base of the device; it is *not* the rectified component of oscillation as in the Class C tube oscillator.

Put another way, solid-state oscillators do not make use of 'grid-leak bias' in the manner of electron tubes. For practical purposes, the solid-state oscillators can operate as a Class A or Class B types, but do not ordinarily work in the true Class C mode. If an attempt is made to bias solid-state oscillators so as to approach Class C operation, it will be found that the oscillator then loses its self-start ability.

On the other hand, solid-state *amplifiers* have more flexibility in this regard than do self-excited oscillators. A solid-state amplifier *can* be biased considerably beyond cut-off, and then be driven with a high-amplitude input signal. The conduction duration will then be quite narrow, as in a 'true' Class C amplifier. Here, unlike the solid-state oscillator, it is fine for the device to be 'dead' between drive pulses.

Using the op amp in oscillators

Of unique interest among active devices suitable for oscillators is the operational amplifier or 'op amp'. This includes various specialized forms of this integrated circuit, such as the voltage comparator, the differential amplifier, the instrumentation amplifier and various power op amps. Demarkation between such IC formats is not always clear-cut and often their roles can be interchanged via appropriate implementation. The op amp is of special interest because it represents the hardware version of the 'universal amplifier' depicted by the triangular symbol in this book. That is, the op amp is a close approach to the idealized active device which, because of its high input impedance, high power gain, low output impedance and basic three-terminal format, can be either used as the active device in an oscillator or can be *symbolically* employed to represent *other* active devices. See Fig. 2.27 for an illustration of this concept.

Thus, the inverting input terminal of the op amp can represent the grid, base or gate or the other active devices; by the same token, the non-inverting terminal would then represent the cathode, emitter or source of those other devices. Finally, the output terminal of the op amp would represent the plate, collector or drain of the alluded devices. Significantly, the depicted op amp oscillator could, at the same time, represent a circuit implementable in its own right. It is assumed that the designer or experimenter knows enough to take care of minor matters such as the unique bias needs of the various active devices, the power-supply requirements, such as dual-polarity d.c. supplies for some op amps and the usual precautions of incorporating by-pass capacitors whether or not they are shown in the circuit diagrams.

The salient difference between the op amp and the various discrete devices is that the op amp comprises a multiplicity of devices, usually involving great complexity. Fortunately, we need not concern ourselves with the complexity, inasmuch as it is neatly built in, and comes to us as a low cost, but sophisticatedly engineered, package. As an example, Fig. 2.7

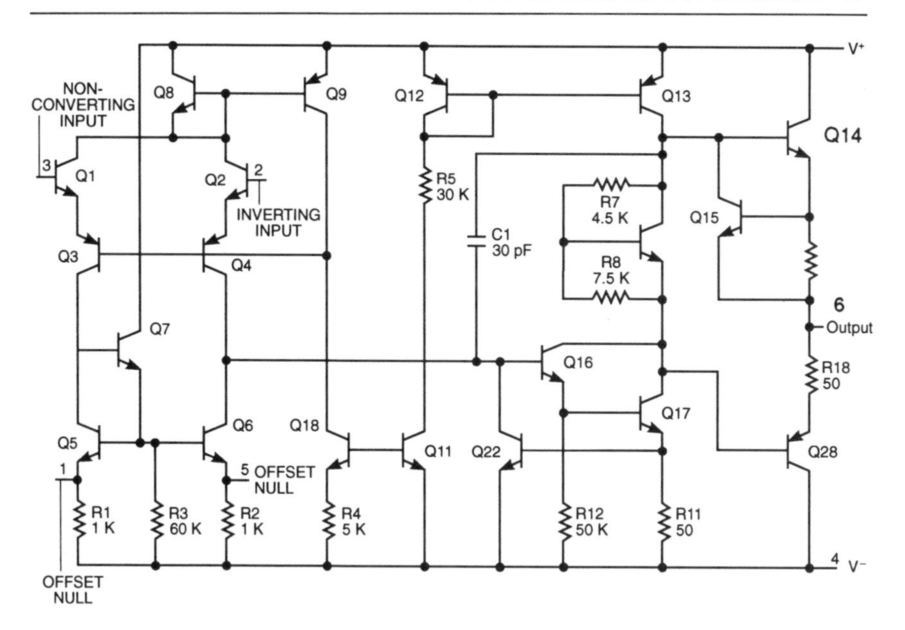

Fig. 2.7 *Schematic diagram of the LM741 op amp. A wide variety of such ICs are available. In addition to low-cost and compact physical dimensions, their salient features are that they can simulate the performance of the discrete devices, and they closely represent the concept of the 'universal amplifier'. Op amps tend to have high input impedance, high gain and low output impedance. This makes them very suitable for numerous oscillator circuits. Frequency capability, power level and temperature behaviour are readily available in manufacturers' literature*

shows the schematic circuit of the long-popular general-purpose LM741 op amp.

Op amps often exhibit more gain than we might wish to use. Excessive gain causes hard-saturating switching action and the resultant square wave can prove difficult to 'tame' with the resonant tank circuit. Also, an op amp operating 'wide open' tends to be vulnerable to latch-up. The simple remedy is to deploy negative feedback. This is implemented by a two-resistor network with the shunt element connected to the output and inverting input terminals and the series element connected to the inverting terminal, as shown in Fig. 2.8. The voltage gain is then very nearly R_2/R_1.

Modes of operation for op amps and logic circuits

Somewhat different considerations prevail for op amps than for discrete devices with regard to oscillator operating mode. An op amp is a composite

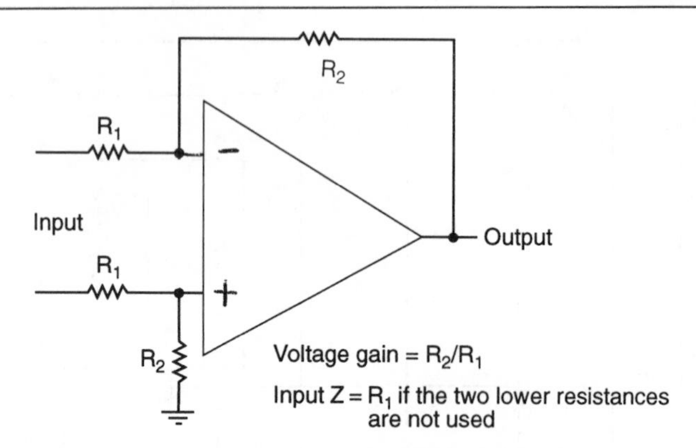

Fig. 2.8 *Negative feedback network for controlling voltage gain of an op amp. The balanced situation shown pertains to an idealized class A amplifier. In most oscillator circuits, the lower two resistances are not needed. Also, many practical oscillators can be implemented with* $R_2 = 100\,k\Omega$ *and* $R_1 = 1\,k\Omega$, *yielding a voltage gain of 100. The LM741 'workhorse' op-amp provides an open-loop voltage gain of about 100 dB at ten 10 Hz. This gain falls at the rate of 20 dB per decade of frequency until it is 0 db at 1 MHz*

device and the amplifier stages are manufactured with their own fixed-bias sources. The reason for this is that the op amp's chief mission is to perform in the linear Class-A mode. So the notion of operation in Class B or Class C is not a practical one for oscillators designed around op amps. Indeed, the op amp merits consideration when, in the interest of oscillator stability and wave purity, exceedingly linear Class-A operation is desired. Inasmuch as most op amps work at relatively low power levels, efficiency and thermal problems have low priority.

Paradoxically, the op amp can *also* function in oscillators as a switching device. As such, it behaves as a more ideal switch than discrete devices operating in their Class B or Class C modes. Such switching action in op amps is termed Class D operation. To a first approximation, it is brought about as the result of overdrive—a situation easily attainable because of the very-high gain in op amps. Indeed, steps are often taken to prevent this mode of operation. On the other hand, other op amp oscillators *deliberately* makes use of the switching mode; the multivibrator is one familiar example. However, an op amp crystal oscillator may also exploit Class-D behaviour, the idea being that the crystal provides the requisite frequency stability. Such circuits are vigorous oscillators and one is not likely to encounter start-up problems or to find that oscillation is readily killed by loading.

The reliability of such over-driven op amp oscillators appeals to many designers. Often a sine wave is not needed. But even when it is, a simple low-pass or band-pass filter or other resonant circuit can convert the square wave to a satisfactory sine wave. In a practical op amp oscillator using, say, a

Hartley circuit, a distorted output wave would probably signify the overdriven mode. In order to obtain a good sine wave, it would be necessary to decrease the amount of positive feedback.

Logic devices operate like overdriven op amps, only 'more so'—they are *deliberately* designed to operate in the switching mode. This is fine for RC and relaxation oscillators of the general multivibrator family. Resonant circuits and crystals are often inserted in the feedback loop to stabilize the frequency, or used as filters to produce a sine wave output. The concepts of Class-A, -B, -C operation have no relevancy here.

Neon bulb as a switching device

The simple two-element neon bulb constitutes a novel switching device. It is but one member of a large family of gaseous diodes, most of which contain an inert gas such as neon, argon, krypton or xenon. When low voltages are applied across the elements of such a device, the gas behaves as a fairly good insulator, permitting passage of virtually no current. As the voltage is raised, the electric force applied to the gas atoms ultimately becomes sufficient to tear outer-orbital electrons from the constraining force of the atomic nucleus. Such disrupted electrons collide with sufficient force with bound electrons to impact these from their respective orbits. The process quickly becomes a cumulative one in which an avalanche of electrons becomes available for collection at the positive element; this migration of electrons constitutes an electric current.

Additionally, atoms from which electrons have been torn have thereby acquired a net positive charge and proceed to the negative element. There they restore, temporarily at least, their lost orbital electrons. This, too, constitutes current and is additive to the total current. The total process, whereby the gas is abruptly changed from an insulator to a conductor is known as ionization. Obviously, this phenomenon can be utilized as a switching mechanism. The important feature of gaseous diodes so far as concerns studies in this chapter, is their voltage hysteresis. This denotes a difference in the voltage causing ionization and the voltage at which the diode deionizes. Specifically we should appreciate that the ionization voltage is higher than the deionization voltage. A gas-diode relaxation oscillator based on voltage hysteresis is shown in Fig. 2.9.

Thyratrons

The gas diode is not a practical switching device for shock-exciting oscillations in a resonant tank circuit because there is no suitable way of synchronizing the pulsing rate with the resonant frequency of the tank. How-

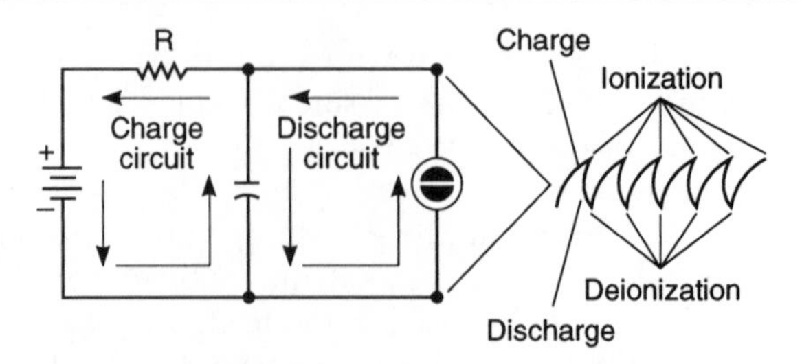

Fig. 2.9 *Gas-diode relaxation oscillator*

ever, this device is useful in conjunction with a charging resistance and a capacitor for generating relaxation type oscillations. The origin and nature of such oscillations are somewhat different from the oscillations derived from tank circuits. A switching device with more desirable characteristics is obtained by the insertion of a control electrode for initiating ionization. Such tubes are made in several different varieties. In the simplest form, a 'cold' cathode is used, that is, simply a metallic rod, not a thermionic emitter of electrons. The thyratron gas tubes are more elaborate, having, among other features, a thermionic emitter that must be supplied with filament current as is the case in vacuum tubes. An improved relaxation oscillator using a thyratron is shown in Fig. 2.10.

Some thyratron tubes contain an additional element, a screen grid that improves the operating characteristics. The chief function of the screen grid is to reduce the control-grid current, which is otherwise appreciable even before cathode-plate conduction has been established. This results in better isolation between input and output circuits. All of these controlled gas tubes suffer from the limitation that once conduction is established between cathode and plate, the control element loses further control over plate current, (Fig. 2.11). Stoppage of plate current is thereafter brought about only by lowering the plate-cathode voltage below the value required to sustain ionization. As a consequence, these tubes also find their most important use in relation oscillators, rather than in oscillators employing resonant tank circuits. The switching rate of all gas tubes is inherently limited by the time required for the gas atoms to deionize. This prevents generation of pulses beyond several tens of kilohertz.

The thyratron inverter

The thyratron inverter is a relaxation oscillator of the multivibrator family. The unique features of this circuit involve the use of thyratron tubes and transformer plate loads. These circuitry features permit operation at much higher power levels than are readily obtained from the conventional multivibrator oscillator. Due to the time required for the gas to deionize, the

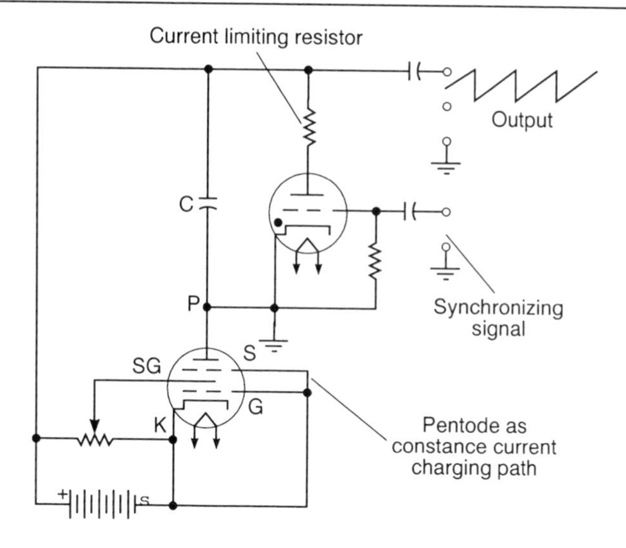

Fig. 2.10 *Improved relaxation oscillator*

frequency of the thyratron inverter is limited to several hundred hertz for mercury vapour tubes and, perhaps, several kilohertz for inert gas tubes. (Ionization time is a small fraction of deionization time and is not an important factor in the matter of frequency limitation.) A thyratron, once fired, cannot be restored to its nonconductive state by variation in grid, or starter electrode, voltage. This imposes a special problem. In order that the two tubes may alternate conduction states in multivibrator fashion, a circuit provision must be made for extinguishing the tubes by means other than the grid signal.

We recall that a conductive gas tube may be deionized by depressing, or removing, its anode voltage. This technique is employed in the thyratron inverter. Capacitor C_x (Fig. 2.12), the so-called commutating capacitor, transfers a negative transient from the plate of the tube switching into its conductive state to the plate of the alternate tube, thereby causing deionization in the latter tube. In this way, the tube that in a multivibrator would be turned off by a grid signal is here turned off by momentary depression of plate potential below the value required to sustain ionization. On the other hand, each tube is alternately turned on in response to grid signals transferred via the cross-over feedback paths, as in the multivibrator.

The frequency of the thyratron inverter is quite sensitive to load variations. This is so because the time constant of the inductive plate circuits is affected by the effective load resistance reflected back into the primary winding of the transformer. A load with an appreciable reactive component can neutralize the effect of the commutating capacitor, thereby stopping

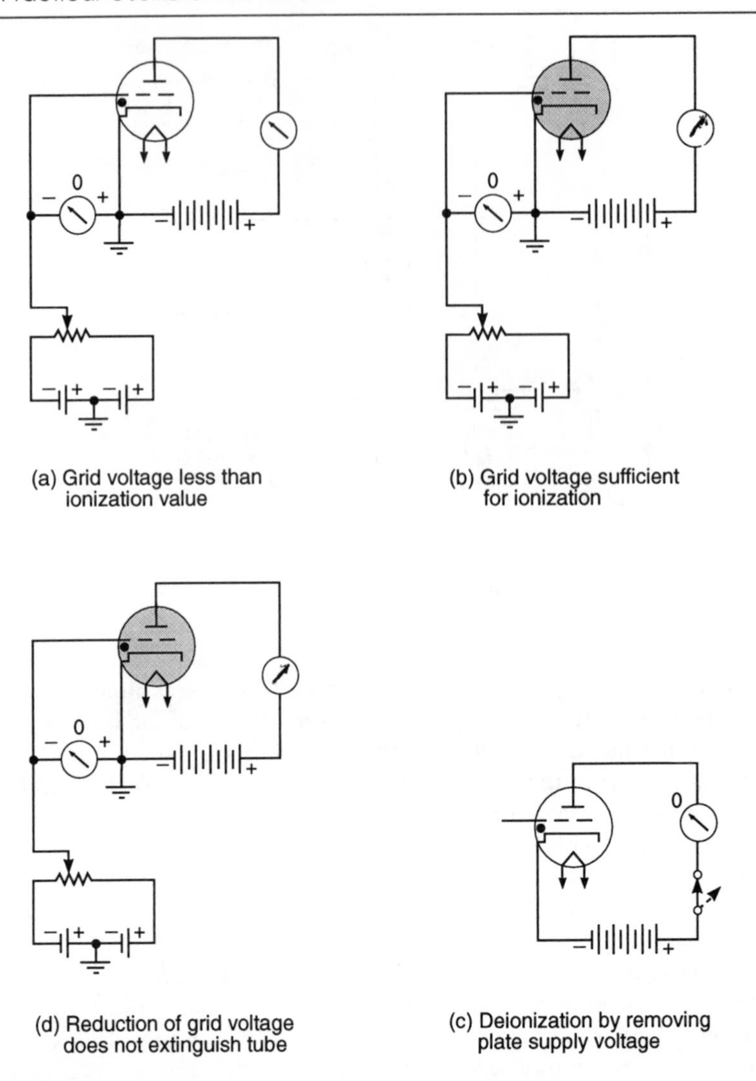

Fig. 2.11 *Control of gas tubes*

oscillation. The output waveform tends to be an approximate square wave. A rough approach to a sine wave may sometimes be attained by inserting a choke in series with one of the power supply leads. This slows the rise and decay of the switching cycles. Contrary to the appearance of the circuit, there is virtually no filtering action due to the 'tank' circuit formed by the commutating capacitor in conjunction with the primary winding of the transformer.

The resonant frequency of this LC combination is generally much higher

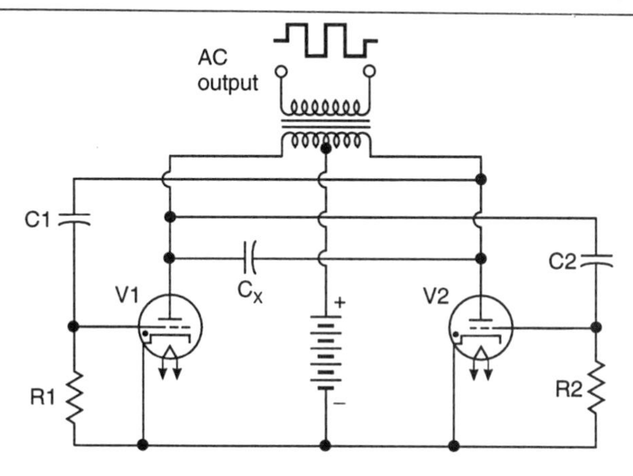

Fig. 2.12 *Basic thyratron inverter*

than the switching rate of the thyratrons. Even if the inverter pulsed at the natural resonant frequency of this LC combination, the effective Q of the tank would be too low to provide much frequency selectivity. This is due to the fact that one half of the transformer primary is always short-circuited. At the power levels generally used in electronics, thyratron inverters are not as efficient as, nor do they possess other desirable performance characteristics readily obtainable from, transitor saturable-core oscillators. However, thyratron inverters have potential application in power systems where it is desired to transform sizeable amounts of d.c. power to a.c. power. The tubes used for high power conversion resemble ignitrons or mercury-arc tubes with starter electrodes. These tubes do not have thermionic emitters in the ordinary sense, but can, for practical purposes, be considered as big brothers to the thyratron.

Spark-gap oscillator

If we connect a high-voltage source of a.c. across a gap formed by two metal electrodes separated a small fraction of an inch, a noisy display of sparks is produced in the gap. Although the flashing display of the dielectric breakdown of the air in the gap appears to the eye as a continuous process, such is not actually the case. Rather, the arc is extinguished each time the voltage impressed across the gap passes through zero (or a relatively low value). For the a.c. sine wave, this occurs at double the rate of the frequency (Fig. 2.13). Thus, an arc energized by 60 Hz high voltage goes out 120 times per second. We see that such a spark gap is effectively a switch that interrupts a high voltage source 120 times/sec. If properly associated with a resonant LC circuit, these disturbances can provoke oscillations at the resonant frequency

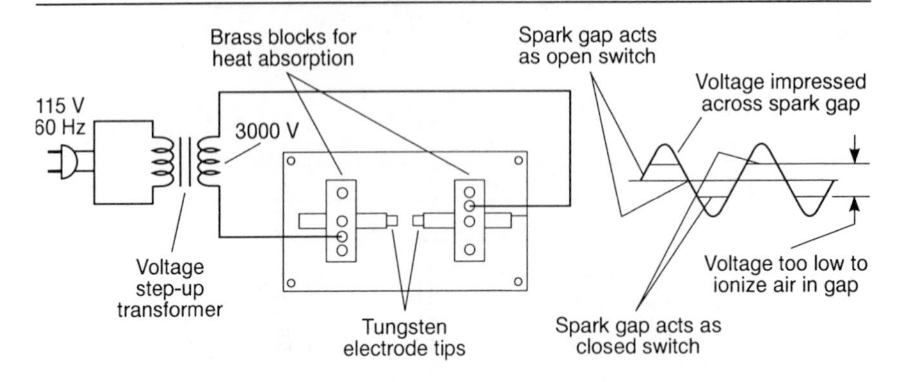

Fig. 2.13 *The spark gap as a switching device*

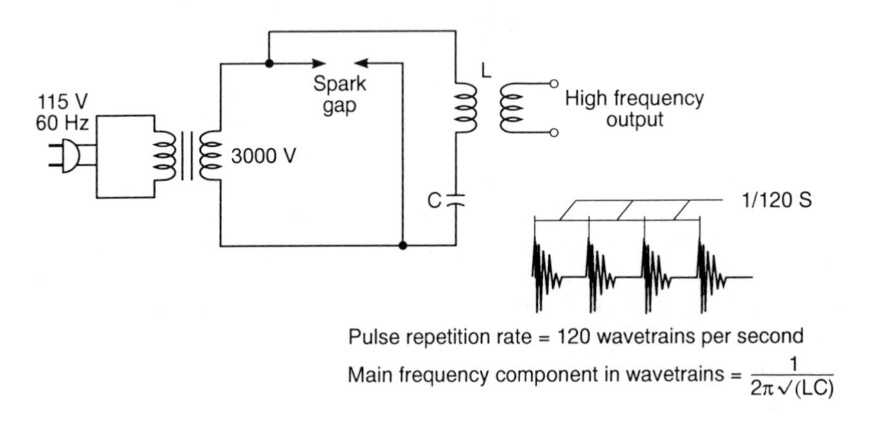

Fig. 2.14 *A spark-gap oscillator*

of the LC circuit (Fig. 2.14). The step-up transformers are special high leakage-reactance types.

The LC circuit must produce many oscillation cycles for each switching pulse. Therefore, the oscillatory wavetrain is highly damped and generally decays to zero long before excitation by the forthcoming switching pulse. A considerable amount of the conversion of low-frequency to high-frequency power in the spark-gap oscillator is invested in harmonics and spurious frequencies that are not readily attenuated by practical LC tuned circuits. For this reason, the spark-gap oscillator is of historical interest only so far as concerns its use in radio communications. However, spark-gap oscillators remain useful for medical and industrial applications. For example, the spark-gap oscillator is often employed in conjunction with welding machines for initiating the arc between the welding rod and the work material. (Once the air between the welding rod and the work material is ionized by the high-voltage, high-frequency energy from the spark-gap oscillator, the low-voltage, low-frequency supply from the welder is able to establish the heavy current arc required to generate welding heat.)

Negative-resistance devices

A family of important oscillation-provoking devices is represented by the so-called negative-resistance elements. Although feedback oscillators, as well as relaxation, phase shift and spark-gap oscillators can be shown, by mathematical analysis, to be amenable to treatment with the concept of negative resistance, such oscillators are not, in the most direct and practical sense, of the negative-resistance variety. We say that an element displays negative resistance in that region of its voltage-current characteristic throughout which a decrease in voltage across the element produces an increase in current through the element, or throughout which a decrease in current through the element produces an increase in voltage across the element. We see that such relationships are contrary to what we would expect from ordinary Ohm's law conduction. When such an element is operated in parallel with an LC tank circuit, it is actually possible to cancel the dissipative losses in the tank.

Sustained oscillations build up in an LC tank when it is associated with the appropriate negative resistance element. We see, that in both the mathematical and the practical sense, the term *negative resistance* is quite descriptive. Looking at the phenomenon from another angle, we can say that if ordinary or *positive resistance* dissipates power, our negative resistance must provide power. This it does. The negative resistance element derives its power from a d.c. source and yields a portion of this power to the LC tank. We should appreciate that the negative resistance exists as such for the a.c. oscillations, not for the static d.c. voltage and current which define the operating point of the negative resistance element.

Concept of dynamic resistance

In negative-resistance devices, the negative resistance is not of the static variety such as is obtained by measuring the resistance with a d.c. ohmmeter. Rather, it is a dynamic parameter and exists only for a.c. As such, it cannot be measured at a single operating point. The a.c., for dynamic resistance, is derived by dividing a small change in voltage by the corresponding change in current. If one of these factors represents an increase and the other a decrease, the dynamic resistance is negative. In a resistor, a small increase in applied voltage would produce an increase in current; consequently, the dynamic resistance of a resistor is positive. In the dynatron tube, a negative-resistance device that is discussed in subsequent paragraphs, an increase of plate voltage from 20 to 25 V might be accompanied by a plate-current change of 2.5 to 1.5 mA. This, we compute as 5 V divided by–1 mA or a negative resistance of 5000 Ω.

The dynatron oscillator

The dynatron oscillator makes use of an interesting vacuum-tube characteristic for provoking oscillation in an LC tank circuit. Certain screen-grid tubes, particularly the older Types 22 and 24 tubes, exhibit a plate-current versus plate-voltage relationship such as that depicted in Fig. 2.15. As plate voltage is increased from zero, we note that the plate current increases up to a certain plate-voltage value. If the plate voltage is increased further, we see that the resultant plate current decreases not merely to zero, but actually to a negative value. Over the region that this inverse current–voltage relationship occurs, the slope of the graph is opposite to those portions of the graph that depict an increase of current with a voltage increase.

The existence of the negative-resistance region results from the simple fact that throughout this region an increase in plate voltage results in a decrease in plate current. Mathematically, when plate current is caused to change from, say, a small positive value to a small negative value, the change is in the nature of a decrease. With different control-grid and screen-grid voltages, good dynatron operation can be obtained with plate currents always positive; that is, with the plate always *consuming* current from the d.c. plate power supply. (This does not conflict with the fact that the dynatron *delivers* a.c. power to the resonant LC circuit.) It should be appreciated that in ordinary Class-A amplification, the polarizing and signal potentials are such that the tube operates within the confines of the near-horizontal region of its plate-current/plate-voltage characteristic. In this way, amplification is maintained fairly linear, and the oscillation provoking tendency of the negative-resistance region does not cause trouble. But, why does this negative-resistance region exist?

Secondary emission

In a vacuum tube, we know that thermionically emitted electrons from the cathode are attracted to other tube electrodes that are biased positive with respect to the cathode. The movement of such electrons within the tube constitutes the flow of electric current and is so indicated by a current meter inserted in a positive grid or plate lead. However, thermionic emission is not the only process whereby free electrons can be generated within a vacuum tube. In a simple vacuum-tube diode, we know that increased plate voltage results in increased plate current (assuming operation is below plate-current saturation). The increased plate voltage not only attracts more electrons per unit time from the cathode area, but imparts greater acceleration to these

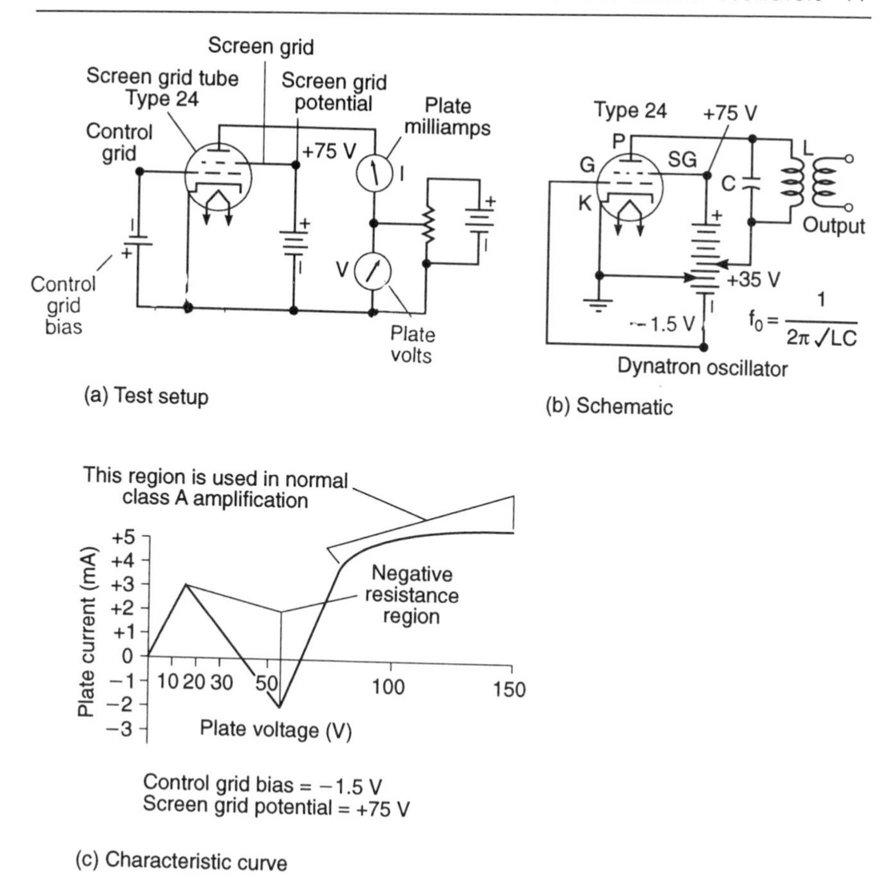

Fig. 2.15 *Negative resistance in the dynatron circuit*

electrons. If the plate voltage is high enough, the attracted electrons have sufficient kinetic energy to impact orbital electrons from the surface atoms of the plate metal. Electrons thus freed from their atomic bonds are called secondary electrons to distinguish them from the primary electron emission from the heated cathode. (No appreciable elevation of plate temperature is required for secondary emission.) Significantly, a single primary electron can liberate two or more secondary electrons. In the diode, or Class-A triode, secondary emission from the surface of the plate is not of any great consequence because such electrons are simply attracted back to the positive plate. Let us investigate what can happen in certain tetrodes.

Reason for dynatron property of negative resistance

Suppose that considerable secondary emission is being produced at the surface of the plate in a tetrode or screen-grid tube, and that the screen-grid voltage is higher than the plate voltage. We see that it is only natural that the secondary electrons accelerate to the screen grid rather than return to the plate. This being the case, screen-grid current must increase at the expense of plate current. (Secondary electrons which leave the plate constitute a flow of current away from the plate.) The higher the plate voltage, to an extent, the more pronounced is the generation of secondary electrons. We have the condition wherein increased plate voltage produces a decrease in plate current. This condition manifests itself as negative resistance when the tube is properly associated with an LC tank circuit. At low plate voltages, insufficient secondary emission occurs to produce the negative-resistance characteristic. At plate voltages approaching and exceeding the value of the positive screen-grid voltage, the secondary electrons return in very greater numbers to the plate. These two conditions establish the limits of the negative-resistance region.

We see that only a small portion of the plate-current versus plate-voltage relationship is suited for dynatron operation. Unfortunately, modern versions of the Types 22 and 24 tubes are manufactured under a process designed to inhibit generation of secondary electrons. Pentode type tubes have a so-called suppressor grid located between the screen grid and the plate. This added element is operated at cathode potential. Electrons liberated from the plate bear a negative charge as to the primary electrons emitted from the cathode. Therefore, the negative suppressor grid repels the secondary electrons, causing them to return to the plate, rather than accelerate to the screen grid. This results in the pentode tube having a more extensive region over which linear amplification can be obtained, but it is obvious that in the pentode we have eliminated the very mechanism required for dynatron negative resistance.

Transitron oscillator

Paradoxically, although the pentode will not perform as a dynatron oscillator, it has more or less replaced the tetrode dynatron tube as a negative-resistance device for provoking oscillation in an LC tank circuit. One of the disadvantages of the dynatron is the dependency upon secondary emission. This varies from tube to tube and with ageing. Indeed, it is quite difficult to obtain satisfactory dynatron action with present day tetrodes, for these are processed during manufacture to make secondary emission less copious than in tubes of earlier production. The pentode can be connected as a transitron oscillator to provide a negative-resistance region which, insofar as concerns the resonant tank, behaves very similarly to the dynatron. The essential difference between the two circuits is that negative resistance in the dynatron involves the relationship between plate voltage and plate current, whereas negative resistance in the transitron involves the relationship between screen-grid voltage and screen-grid current.

Reason for transitron property of negative resistance

In Fig. 2.16a we see that the suppressor grid of a pentode connected to exhibit transitron characteristics is biased negative with respect to the cathode. However, the actual value of the negative suppressor voltage is determined by the screen-grid voltage. This is so simply because the screen-grid voltage supply and the suppressor bias supply are connected in series. If we increase the screen-grid voltage, the suppressor grid bias becomes relatively less negative. This being the case, electrons passing through the screen-grid apertures are less likely to be repelled back to the screen grid; rather, such electrons are more likely to reach the plate where they add to plate current, but at the expense of screen-grid current. The converse situation likewise operates to establish the negative-resistance region. That is, if screen-grid voltage is decreased, suppressor-grid bias becomes relatively more negative, and electrons passing through the screen grid are deflected back to the screen grid, thereby adding to screen-grid current at the expense of plate current.

Thus, we see that screen-grid current changes are opposite in direction to screen-grid voltage changes. Such a negative resistance does not exist over the entire range of screen-grid current versus screen-grid voltage characteristics, but only over a small portion of the total characteristics. We see this must be so because suppressor voltage cannot become positive, for then it would draw current as the result of attracting electrons that otherwise would be collected by the plate. At the other extreme, suppressor voltage cannot become too negative, for then it would cut off plate current by deflecting electrons back to the screen grid. For the transitron action to exist, the plate must be allowed to participate in apportioning the total space current. This it can do only when suppressor voltage is neither too low nor too high.

Function of capacitor in transitron oscillator

The capacitor, C_2, in the transitron oscillator circuit of Fig. 2.16b, communicates the instantaneous voltage developed across the tank circuit to the suppressor, where it is superimposed upon the fixed negative bias voltage. It is educative to observe that the suppressor grid exerts stronger control within the negative-resistance region than does the screen grid. Thus, when the voltage impressed at these two elements increases, we expect screen grid current to increase, for such would be the case at an anode, which the screen

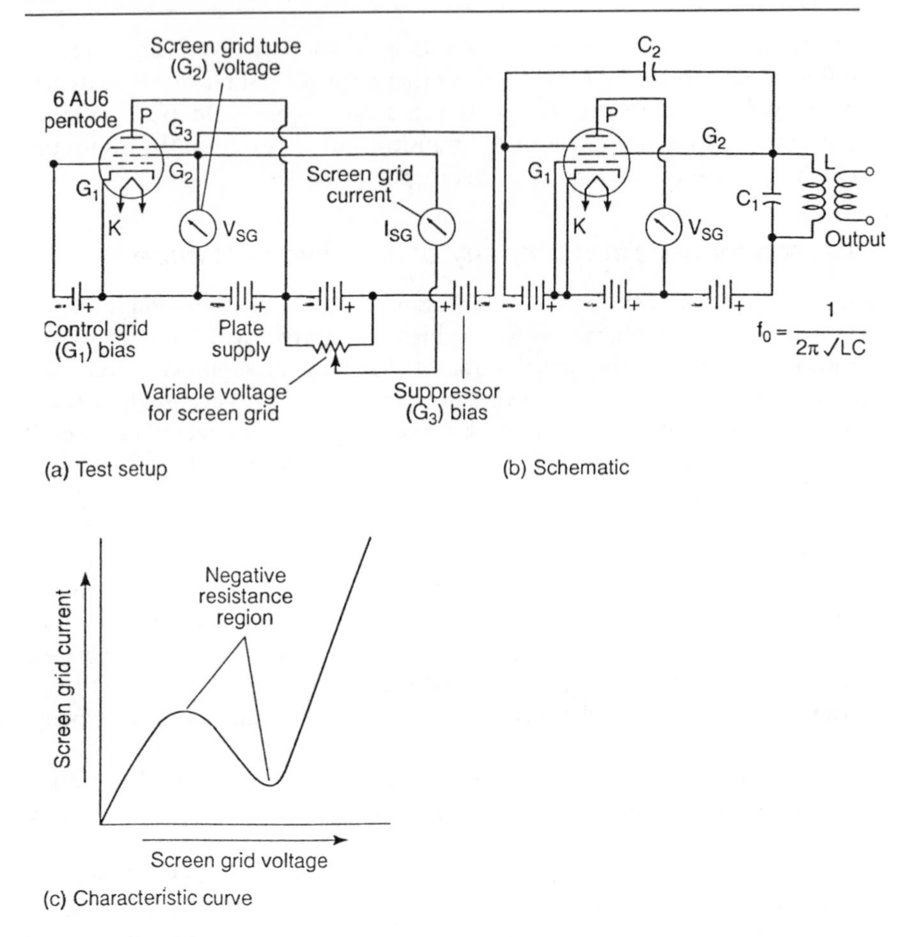

Fig. 2.16 *Negative resistance in the transitron circuit*

grid simulates. However, the simultaneous decrease in negative voltage applied to the suppressor has a more pronounced effect upon the screen-grid current. As explained, the additional electrons that are now allowed to pass to the plate would in ordinary circuits have been collected by the screen grid. Raising the voltage at the screen grid in a transitron circuit causes a decrease in screen-grid current. Like the dynatron, the transitron is a two-terminal oscillator, it being only necessary to connect the two terminals of a parallel-tuned LC circuit. Some of the tubes that make satisfactory transitron oscillators are the 6AU6, 6BA6, 6J7, and 6K7.

Oscillation occurs when the dynamic screen-grid resistance is numerically equal, but of opposite sign to the equivalent resistance of the resonant LC tank; that is:

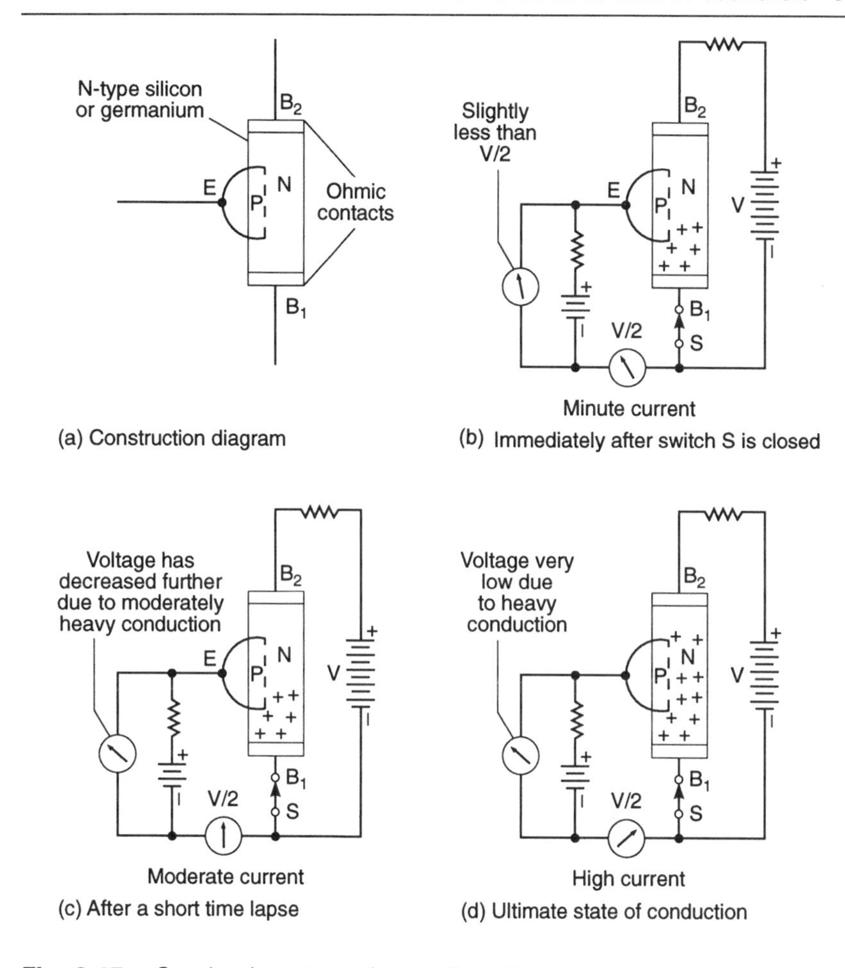

Fig. 2.17 *Conduction states in a unijunction transistor*

$$r_{SG} = -\frac{L}{(R_S)\ (C)}$$

R_S, as we recall, is the total series resistance of L and C.

The unijunction transistor

The unijunction transistor is a semiconductor switching device with characteristics somewhat suggestive of thyratron action. The physical configuration of the unijunction transistor is shown in Fig. 2.17a. It is essentially a pn structure in which two base connections, rather than the single base connection of ordinary rectifying junction are provided. We know that reversal of applied polarity changes conduction from a high to low value in the ordinary junction diode. In the unijunction transistor, which is essentially a double-base junction diode, reversal of polarity between the centrally located p, or emitter element, and the n-type bar occurs abruptly, giving rise to a switching process in the external circuit. Consider the situation depicted in Fig. 2.17b. When the switch is initially closed, the p element will assume a potential relative to base No. 1, B_1, equal to approximately half of the voltage applied by the battery connected to the two bases. This is true because the resistivity of the semiconductor bar is uniform along its length. At first, the major portion of the p element will be substantially in the nonconducting state because of insufficient voltage gradient between it and the n material of the semiconductor bar. However, the lower tip of the p element will be slightly in its forward-conduction region, that is, the lower tip will act as an emitter of positive charges (holes). (This is because the lower tip is off-centre.)

The initial conductive state following closure of switch S is not stable. The positive charges injected from the lower tip of the p element rapidly 'poison' the lower half of the semiconductor bar. These charges are collected by negatively polarized base No. 1. The movement of these charges through the lower half of the bar constitutes current. In essence, the conductivity of the lower half of the bar is increased. This, in turn, shifts the distribution of potential drop within the bar so that the entire p element finds itself relatively positive with respect to adjoining n material of the bar. As a consequence, the entire p element becomes an injector of positive charges, thereby increasing conductivity of the lower half of the bar to a high degree. This occurs with extreme rapidity.

This is the stable end result of the circuit as shown in Fig. 2.17d. The p element, that is, the emitter, in conjunction with the n material of the lower half of the bar and the ohmic base connection (No. 1), behaves as an ordinary pn junction diode passing heavy current under the operation condition of forward conduction.

Instead of the battery, $V/2$, let us substitute a capacitor and a charging resistor (see Fig. 2.18). For a time, the capacitor will charge much as it would if the p element and base No. 1 acted as a very high resistance. When the capacitor voltage attains the value F (approximately one half the base-to-base voltage), depicted in the curve of Fig. 2.19, the cumulative process of increased charge injection and spread of charge-emitting surface will be initiated. The capacitor will then 'see' a relatively low resistance and its stored voltage will be rapidly depleted. During this discharge, the avalanching of charge carriers continues within the lower half of the bar until the capacitor voltage is quite low–in the vicinity of 2 V. In other words, insofar as concerns the external circuit connected between the p element and base No. 1, there is a voltage-hysteresis effect analogous to that of the neon-bulb relaxation oscillator wherein ionizing voltage is higher than extinction

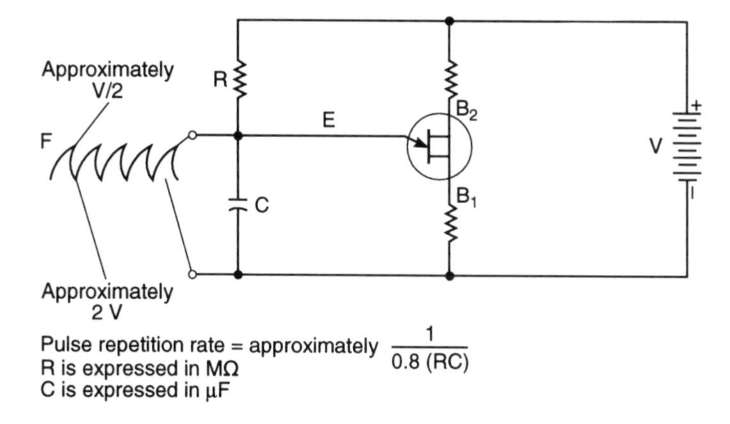

Fig. 2.18 *Basic unijunction relaxation oscillator*

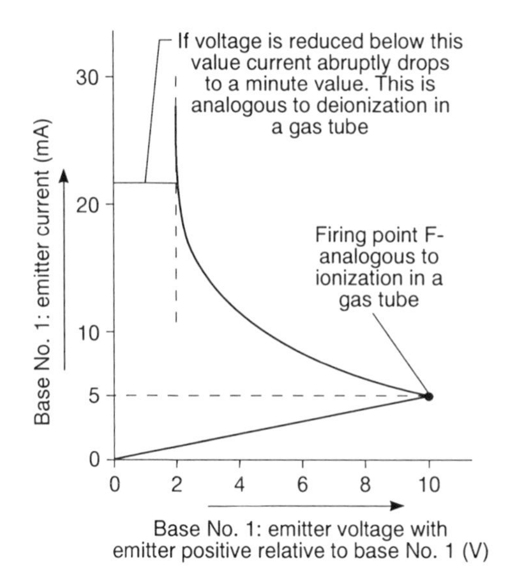

Fig. 2.19 *Current-voltage relation in Base No. 1-to-emitter diode*

voltage. In any event, when the capacitor discharges to a sufficiently low voltage, injection of positive charges into the bar ceases, and the lower half of the bar reverts to its initial state of high resistivity. The capacitor charge-discharge cycle then repeats. In some cases the capacitor-charging resistor may be dispensed with. The capacitor then receives its charging current through the reverse-conducting upper half of the bar, which acts as a constant-current generator. This causes the capacitor to charge linearly with

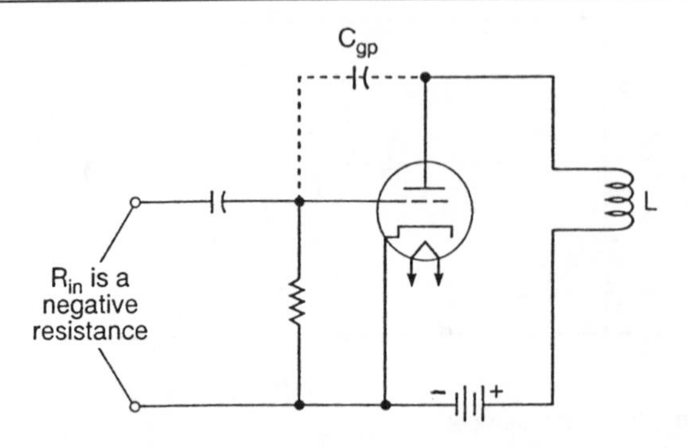

Fig. 2.20 *Under certain conditions the input circuit of a triode may appear as a negative resistance*

respect to time and the resultant sawtooth has a suitable shape for many sweep circuit applications. This mode of operation imposes the practical limitation that the pulse repetition rate cannot be made as high as when a charging resistor is used.

Triode input as negative resistance

There is an important circuitry family of triode-tube oscillators, the operation of which is generally ascribed as being due to feedback via the internal plate-grid capacitance of the tube. Actually, such oscillators are more closely related to those of the negative-resistance family than to the feedback group. This is so despite the fact that it can be shown mathematically than even 'true' feedback oscillators present negative resistance to their resonant tanks. In those oscillators we classify as of the feedback variety, the negative-resistance property is intimately associated with the presence of the resonant tank. However, in the so-called negative-resistance oscillator, the negative-resistance property is displayed by the tube or other device with or without the resonant tank.

These considerations bring us to the interesting fact that the grid-cathode circuit of a 'grounded-cathode' triode amplifier can be made to appear as a negative resistance. This is brought about by making the plate load inductive (see Fig. 2.20). For a certain range of plate-circuit inductance, the grid circuit is capable of supplying power to a resonant tank. The action involved here is not easy to visualize without considerable mathematical analysis. Nevertheless, the effect is an important one and is a basic principle in the operation of the tuned-plate tuned-grid oscillator, certain variations of the Hartley oscillator, and the Miller crystal oscillator. These oscillators will be considered in detail in our discussion of practical devices. Although useful negative input resistance does not appear for just any value of plate circuit inductance, it is found in practice that a broad range of inductance values suffice to bring about useful values of negative input resistance. This is particularly true when the Q of the inductive plate load is high. Throughout the range of inductance over which the input circuit contains negative resistance, the effect of varying the inductance is to change the numerical value of the negative input resistance. The value of the input resistance can be found with the following formula:

$$R_{in} = -\frac{1}{2\pi f C_{gp} K \sin\theta}$$

where:

R_{in} is the input resistance in ohms,
C_{gp} is the effective grid-plate capacitance, expressed in farads,
f is the frequency in hertz,
θ is the phase angle of the plate load,
K is the voltage gain of the tube.
Note: θ and $\sin\theta$ are positive for an inductive load. This makes R_{in} negative.

(The input capacitance of the tube also changes with variations in the nature and magnitude of the plate load. This is not relevant to our particular discussion, however.)

The electron tube provides the classic explanation of this phenomenon. The same thing happens, however, with bipolar transistors and with the MOS devices.

The saturable magnetic core

The saturable magnetic core is at once an oscillating element and an oscillation-provoking device. Nevertheless, it requires the association of a controllable switch such as a transistor in order to produce oscillation. The generated waveform is nonsinusoidal; such an oscillating system is of the relaxation type.

The best core material for this type of oscillation exhibits a rectangular hysteresis loop, such as that shown in Fig. 2.21b. Suppose the battery circuit has just been completed. Conduction will start in the emitter-collector diode of one or the other transistors, but not in both because whichever one first conducts induces a reverse bias in the base-emitter diode of the opposite transistor, which prevents conduction. Due to inevitable circuit and transistor imbalances, it is highly probable that such a starting mechanism will follow closure of the battery circuit switch. Sometimes, in order to ensure

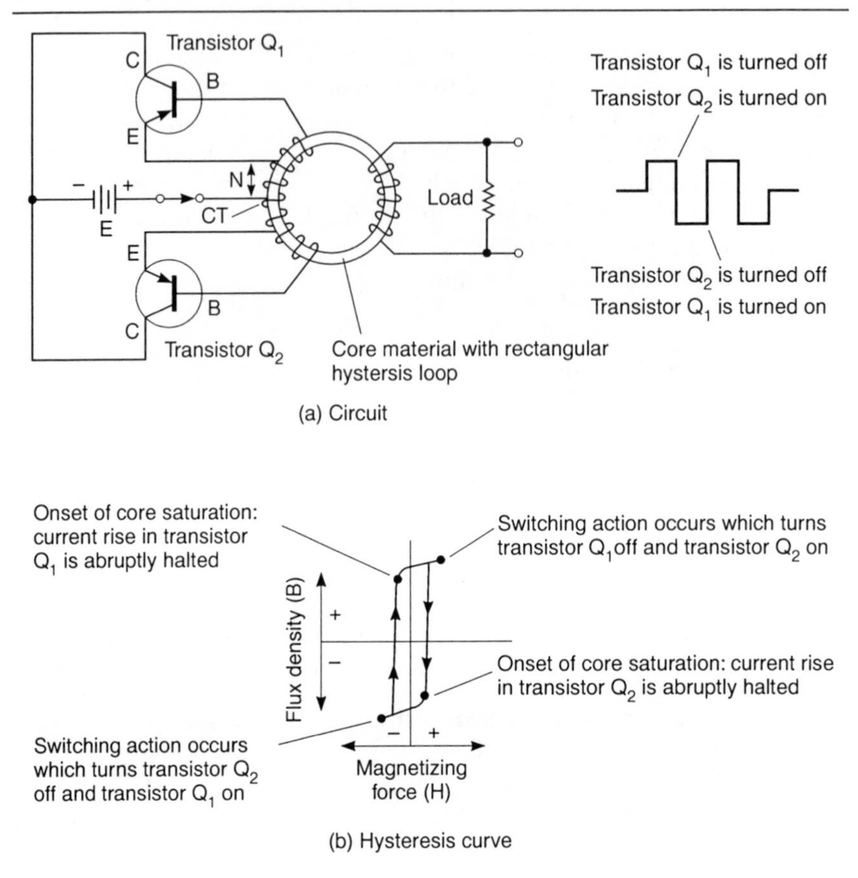

Fig. 2.21 *The saturable core as an oscillation-provoking device*

starting reliability, one transistor is provided with a small amount of forward bias, or each is provided with a small amount of forward bias, or each is provided with different forward biases. Forward bias is readily provided by inserting a small resistance in series with the base of a transistor, and a second, higher resistance between collector and base terminals.

Oscillation in the saturable-core circuit

In any event, we will assume that conduction has commenced in one of the transistors, say the one labeled Q_1 in Fig. 2.21. This conduction results in a rapidly increasing current through the portion of the winding between CT and the emitter of Q_1. As this current increases, it induces *two* important voltages. One of these appears across the portion of the winding between emitter and base of Q_1. This voltage is of such polarity as to drive Q_1 into increasingly heavy conduction. The other voltage appears across the portion of the winding between emitter and base of Q_2 and is of such polarity as to clamp Q_2 in its nonconducting state. The rapidly increasing current through the portion of the winding which feeds the emitter of transistor Q_1 ultimately causes abrupt magnetic saturation of the core. When this occurs, electromagnetic induction of the voltages controlling the bases of the two transistors ceases. However, these voltages cannot instantaneously return to zero due to the energy stored in the magnetization of the core. Accordingly, the conduction states of the two transistors remain for a time as they were immediately preceding the onset of core saturation. However, once the collapse of the magnetic field gets well under way, sufficient reverse-polarity voltages are electromagnetically induced in the base-emitter portions of the winding. This reverses the conduction states of the transistors, turning Q_1 off and Q_2 on. The conduction of the emitter-collector diode of Q_2 now repeats the cycle of events, but with core saturation ultimately being produced in the opposite magnetic sense. After an interval following this second saturation process, the transistors again exchange conduction roles, and the core is on the way to saturation in the same magnetic sense as originally considered. Thus, the transistors alternate states: when Q_1 is on, Q_2 is off and vice versa.

Two additional saturable core oscillators are shown in Fig. 2.22. Saturation flux density, B_s, for various core materials is given in Table 2.1. The following design data applies to Fig. 2.22 and to Fig. 2.21 as well:

$$N = \frac{E \times 10^8}{4f\ B_s\ A}$$

where:

N is the number of turns from CT,
E is the battery voltage,
f is the frequency of oscillation in hertz,
B_s is the flux density at saturation lines per sq. cm,
A is the cross-sectional area of core in sq. cm.

The number of turns in the windings indicated by 'n' in Figs 2.22a and b generally is equal to $N/8$ to $N/3$.

The frequency of oscillation is determined by the battery voltage, the number of turns in the portion of the winding that feeds the emitters of the transistors, the flux density at which saturation occurs, and the cross-sectional area of the core. However, once the oscillator has been constructed, battery voltage is the only frequency-governing parameter that requires consideration, the others being fixed by the number of turns placed on the core, and the nature and geometry of the core material. This type of oscillator is very efficient, efficiencies exceeding 90% being readily obtainable. We note that there is no need for high power dissipation in the transistors. When a transistor is off, its collector voltage is high, but its collector current is

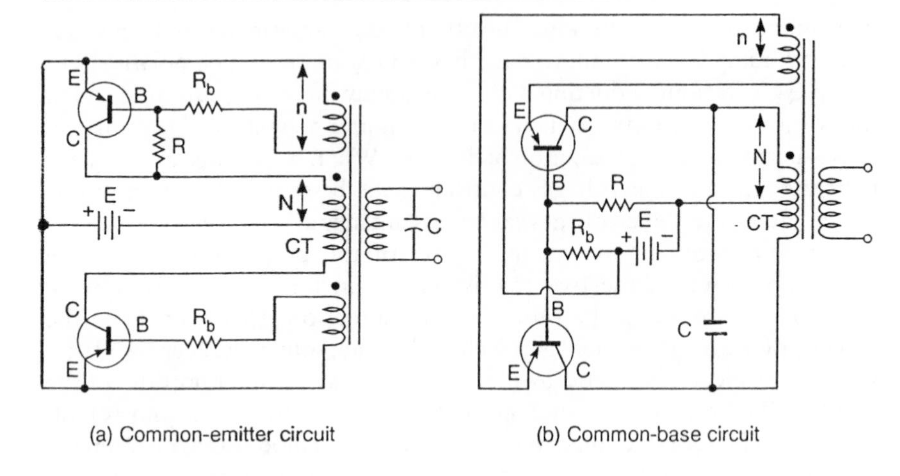

(a) Common-emitter circuit

(b) Common-base circuit

Fig. 2.22 *Saturable-core oscillator circuits*

Table 2.1 *Saturation flux densities,* B_s, *for various core materials**

Core material	Saturation flux density in kilogauss (thousands of lines/sq. cm)
60 Hz power transformer steel	16–20
Hipersil, Siletcron, Corosil, Tranco	19.6
Deltamax, Orthonol, Permenorm	15.5
Permalloy	13.7
Mollypermalloy	8.7
Mumetal	6.6

* When using these B_s values, make certain that the core area, A, is expressed in sq. cm.

virtually zero. Conversely, when a transistor is fully on, its collector current is high, but its collector voltage is very low. In either conduction state, the product of collector voltage and collector current yields low wattage dissipation in the collector-emitter section. We see that one of the advantages of rectangular hysteresis loop material is that a core of such material permits rapid transition between off and on. This is desirable, because when a transistor is neither fully on nor fully off, the product of collector voltage and current can be high. If the transition is not rapid, efficiency will be lowered and the transistors will be in danger of being destroyed by generation of excessive internal heat. Silicon steel such as is used in 60 Hz transformers is often employed to save costs. However, less power output and lower efficiency are the result.

We often see these saturating-core oscillator circuits with a capacitor connected across the load winding, or even across the primary winding.

However, no LC tank is formed by the addition of such capacitors and the frequency is not changed thereby. The capacitor is added in order to absorb voltage transients that can destroy the transistors. These transients, or 'spikes', are due to the abrupt saturation of the core. The frequency of these oscillators is generally limited, at least in units designed for high power, by the frequency capability of the power transistors. However, frequencies on the order of several hundred hertz to around 100 kHz are readily attainable and are well suited for high efficiency and small, light-weight cores.

The electron beam in a vacuum

In most of the electrical oscillating elements and oscillation-provoking devices we have considered, the described action resulted from the controlled or guided flow of electrons in metallic conductors of various geometric configurations. In waveguides and resonant cavities, we found that the electron was somewhat freed from the restrictions ordinarily prevailing in low-frequency ohmic circuits. In these microwave elements, much of the energy associated with the electron manifested itself as electric and magnetic fields external tjo metallic conducting surfaces. Thus, the electron movement, that is, the electric current in the walls of the waveguide, is at once influenced if we insert an obstruction within the aperture of the guide. Such an obstruction disturbs the pattern of the electric and magnetic force fields. We can go a step further than this. We can free the electron from confinement in matter completely, in which case it becomes highly susceptible to manipulation by various force fields. This is most directly accomplished by some arrangement involving a thermionic cathode and a positively biased anode, the basic objective being to create a beam of electrons in a vacuum.

Historically, the X-ray tube has considerable relevance. In this tube, the electrons are imparted tremendous acceleration by the extremely high potential difference anode and cathode. The energy thereby acquired by the electrons is sufficient to produce dislocations in the orbital electrons of the atoms in the anode metal. Inasmuch as mass and compliance are associated with these orbital electrons, a wave motion in the form of an oscillating electromagnetic field is propagated into space. Although the frequencies of these waves are far beyond those generated by the oscillators dealt with in this book, we should realize that the X-ray tube utilizes the basic principle of all modern microwave tubes. This principle is the conversion of the energy in an electron beam to oscillatory power at a desired frequency.

The magnetron

The magnetron, like the X-ray tube, is a vacuum diode in which thermionically freed electrons are imparted high kinetic energy by an intense electric accelerating field. In the X-ray tube, the staccato-like rain of these accelerated electrons bombard the anode at appropriate intervals to sustain electron oscillation between orbits of the anode atoms. In the magnetron, the objective is to excite oscillations in microwave cavities. We cannot do this by the direct action of electrons simply accelerated by high voltage. Rather, we require the passing motions of high energy electrons at the vicinity of a cavity. This must occur at intervals at, or very near to, the natural resonant frequency of the cavity. The electric field of the electron can, under such circumstances, interact with the oscillator electric field of the cavity in such a way that a synchronized excitation is applied to the cavity once per cycle. In the magnetron, this is achieved by causing the electrons in the cathode-anode gap to describe orbital motions. Let us see how this is done.

Motor action on electrons

We know from electrical theory, that a so-called 'motor' action occurs when a current-carrying wire is exposed to a magnetic field. This, indeed is the principle underlying the operation of electric motors. It is well to appreciate that in the electric motor, the repulsive force is not actually exerted upon the physical wires imbedded in the armature slots, but rather upon the moving electrons within the wire. This being true, we are not surprised that the same motor action may be imparted to moving electrons in a vacuum. The magnetron employs a powerful magnet for this very purpose (Fig. 2.23a). The field is oriented perpendicular to the plane of electron emission from the cathode. At low plate voltages, the motor action results in small tight orbits and the electrons do not reach the plate at all. This is attested by the negligible plate current which exists at low plate voltages (see Fig. 2.23b). At higher plate voltages, the magnetic field has less deflective power and the electron orbits are consequently larger.

In fact, some of the electrons simply 'fall' into the metal and are collected as in the ordinary vacuum diode. The collection of such electrons constitutes plate current and is the source of useless plate dissipation which lowers the efficiency of the tube. Other electrons, however, orbit past cavity apertures at just the right time to excite and sustain oscillations in the cavities. These electrons, together with others which make relatively small orbits by virtue of low initial velocities of emission, return to the cathode where their remaining energy is dissipated as heat. Indeed, in high power magnetrons, it is sometimes possible to turn off the heater current once the magnetron has become operative, the energy of such returned electrons being sufficient to maintain the cathode at the temperature required for thermionic emission. It is quite common to reduce the heater current once the magnetron is oscillating. We see that magnetron action is dependent upon synchronization of the orbital cycles of the electrons with the resonant frequency of the cavities. Oscillation is excited and sustained by interaction

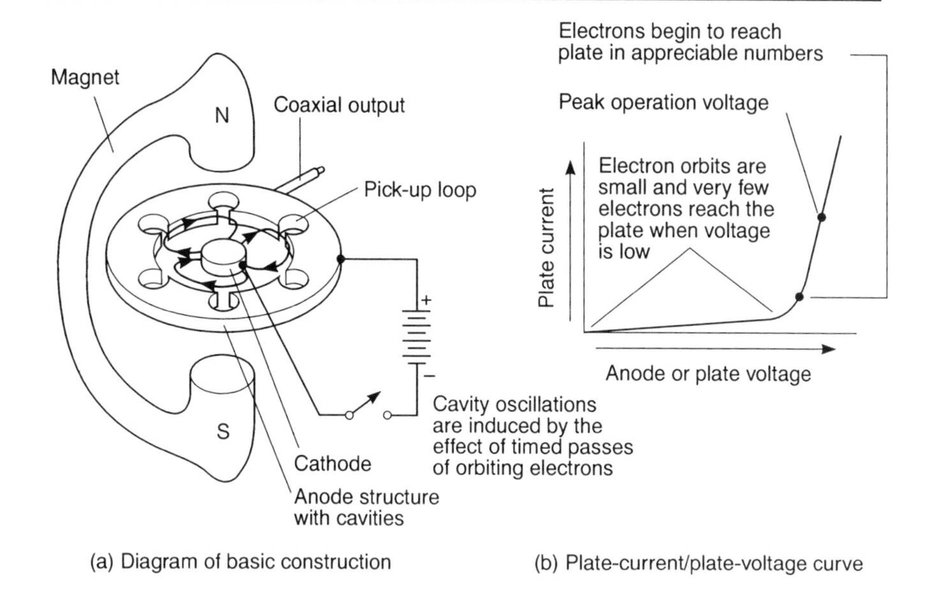

Fig. 2.23 *Multicavity magnetron*

of the electric field in the oscillating cavity and the electric field of the passing electron. It is not necessary or desirable that the electrons be collected by the anode structures. The microwave energy is extracted by a loop inserted in one of the cavities. This couples the total oscillatory energy of all cavities to an external load because all cavities are 'electronically coupled' by the modulating action of the orbiting electrons.

The reflex klystron

The reflex klystron, shown in Fig. 2.24, employs a somewhat different stratagem to extract energy from an electron beam in the form of microwave oscillation. The anode of the klystron is a resonant cavity that contains perforated grids to permit accelerated electrons to pass through and continue their journey. Such electrons are not, however, subsequently collected by a positive electrode. Rather, they are deflected by a negatively polarized 'reflector' and are thereby caused to fall back into the cavity grids. The operational objective of the tube is to have such electrons return to the cavity grids at just the right time to reinforce the electric oscillatory field appearing across these grids. When this situation exists, oscillations are excited and sustained in the cavity. Microwave power is coupled out of the cavity by means of a loop if coaxial cable is used, or simply through an appropriate aperture if a waveguide is used for delivering the power to the

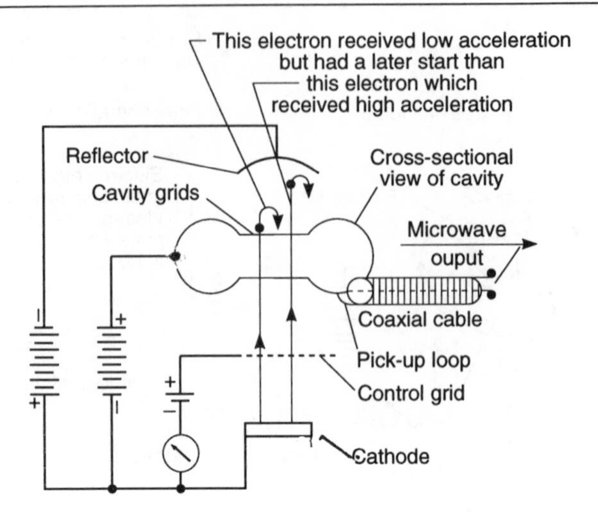

Fig. 2.24 *Basis for oscillation in the reflex klystron*

load. After the kinetic energy of the electrons has been given up to the oscillatory field of the cavity, the spent electrons fall back to the positive-biased control grid where they are collected, thereby adding to control grid current. If the tube is not oscillating, a relatively high number of electrons are deflected by the retarding field of the reflector with sufficient energy to pass through the cavity grids, thence to be collected by the control grid. However, when oscillations are sustained in the cavity, the falling electrons yield most of their energy to the oscillating electric field appearing across the cavity grids. Such electrons are subsequently collected by the cavity grids, which in this function behave as the plate of an ordinary diode. Inasmuch as the spent electrons do not fall into the positive field of the control grid, a profound dip in control-grid current accompanies the onset of oscillation within the cavity.

Bunching of electrons returned to cavity grids

Depending upon their time of arrival from the cathode, different electrons will be imparted different accelerations if oscillations exist in the cavity. Some electrons will experience violent attractions when the oscillating electric field of the lower cavity grid is at its positive crest. Others will be subjected to less intense attractive force because they approach the lower cavity grid at a time of the oscillatory cycle when it is not so highly positive. (During oscillation, the d.c. voltage level of the cavity grid is effectively modulated at the oscillatory rate.) As a consequence, some electrons will be projected higher into the retarding field of the reflector than others. Nevertheless, electrons projected upwards through the cavity grids can be caused to return to the cavity in groups or 'bunches' by adjustment of the reflector voltage. In order for electrons with different penetrating distances into the retarding field to simultaneously return to the cavity grids, such electrons must obviously possess different velocities. The electrons in the region between the reflector and the upper cavity grid are said to be *velocity modulated.* When, under influence of proper reflector voltage, bunches of such electrons return to the cavity grids, the energy they acquired during their fall from the retarding field is extracted by the oscillating electric field across the cavity grids. We see a cumulative process: oscillation produces velocity modulation, and the bunching due to velocity modulation delivers synchronized energy to the cavity grids, thereby sustaining oscillation. In the reflex klystron, as in the magnetron, microwave oscillation is provoked in cavities by the influence of timed electrons. In both instances, the oscillatory power is derived from the energy of electrons accelerated in a vacuum.

Travelling-wave tubes and the backward-wave oscillator

A third method of extracting energy from an electron beam to produce microwave oscillatory power is utilized in the backward-wave oscillator. This device is a specialized member of a family of microwave tubes known as travelling-wave tubes. We will discuss amplification in the travelling-wave tube because this is the basic process that must be understood in the operation of the backward-wave oscillator. In the travelling-wave tube, a beam of electrons is projected through the centre of a long helix (see Fig. 2.25). It is desirable that the electrons should pass close to, but should not be collected by, the positively biased helix structure. This requires formation of an electron beam of very small diameter. This is not, in itself, for an electron beam tends to diverge due to mutual repulsion of the electrons comprising the beam. In order to constrain these electrons within the inner diameter of the helix, a strong longitudinal magnetic field is applied parallel to the axis of the helix. The motor-action experienced by the divergent electrons interacting with this field imparts forces urging them toward the geometric axis of the helix. Such focusing of the electron beam is the first requisite of all travelling-wave tubes. Optimum beam focusing is accomplished by the alignment of the magnet structure, which results in minimum helix current. Let us, now, consider the function of the helix.

Function of helix as 'slow-wave' structure

The helix behaves essentially as an open wave guide. Although not commonly employed in microwave or VHF practice, a single conductor can guide the progress of the electromagnetic waves. The single wire lead-in for

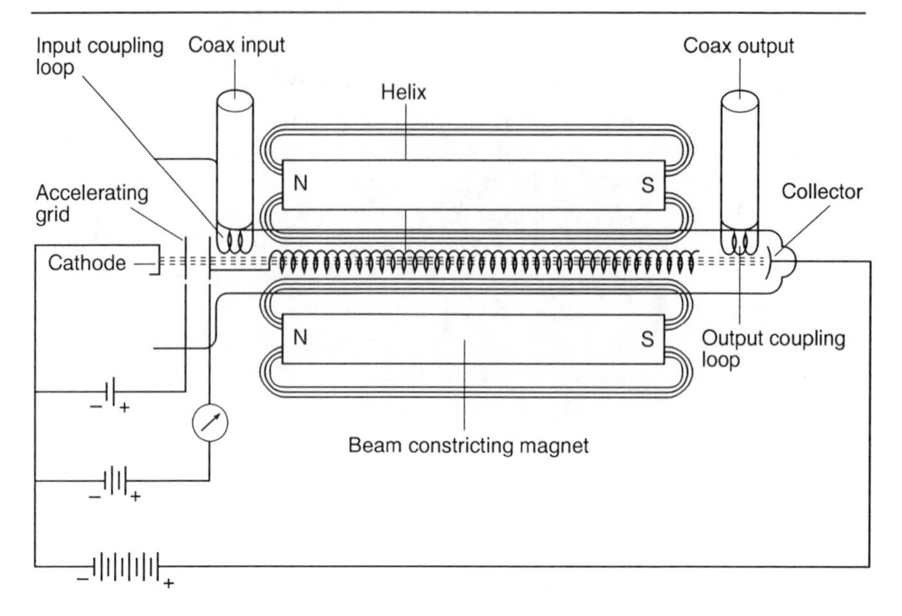

Fig. 2.25 *Travelling-wave tube*

television antennas (the so-called G-string transmission line) is a wave guide of this nature. Despite the fact that a signal may propagate along the surface of the helix conductor at approximately the speed of light, the progress of the waves along the axis of the helix will be only a fraction of this speed, being governed by the winding pitch of the helix. Of prime importance, the speed of the electrons within the beam can be made nearly the same as the speed of the signal waves travelling axially down the length of the helix. When this condition prevails, we have an interaction between the electron beam and the signal. This interaction is such that the signal increases its power level during its axial progress from the cathode end to the collector end of the helix. This power is extracted from the electron beam. At the output end of the tube, the electrons are collected by a positively polarized electrode and the amplified signal is coupled from the end of the helix to a coaxial fitting or to a waveguide aperture. In such a travelling-wave amplifier, it is important that the amplified signal is not reflected back to the input end of the helix in such a way as to promote the build-up of oscillation. This is prevented by making at least a portion of the helix 'lossy'. Reflected waves are then attenuated. Of course, forward-travelling waves suffer attenuation too, but it will be shown that these waves grow in amplitude by absorbing energy from the electron beam.

The travelling-wave amplifier makes an excellent oscillator if a microwave cavity is externally inserted in a feedback loop between output and input. The cavity then determines the frequency of oscillation. The travel-

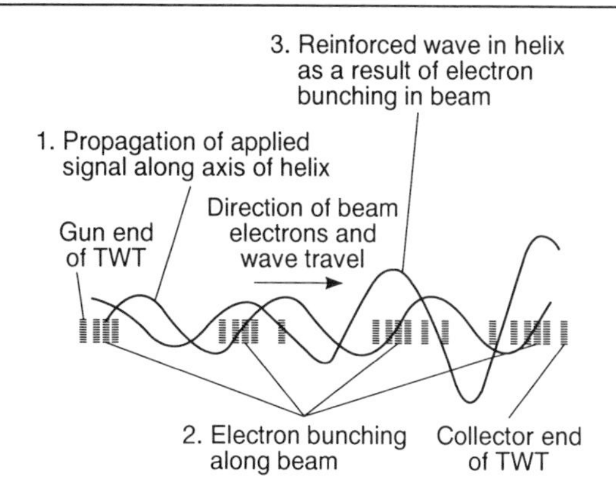

Fig. 2.26 *Wave generation in a travelling-wave tube*

ling-wave amplifier is relatively insensitive to frequency, this being its much heralded feature. This is because the helix behaves much like a transmission line terminated in its own characteristic impedance, i.e., it is nonresonant. Travelling-wave amplifiers commonly cover frequency ranges in the vicinity of two to one. Often the limiting factor to the extent of frequency response is the input and output coupling mediums. (There is considerable practical difficulty encountered in preventing coupling helixes, loops or probes from becoming reactive when operation over extremely wide frequency ranges is attempted.) Of course, we wish to know what makes the axial component of the helix wave grow as it journeys adjacent to the electron beam down the tube.

Bunching of electrons in beam

Referring to Fig. 2.26, we see that the signal wave in its propagation along the helix reacts upon the electron beam. Specifically, the beam is alternately speeded up and slowed down; we say it is 'bunched'. When the electric field of the signal wave increases in the positive direction, the electrons in the beam are speeded up. When the electric field of the signal wave increases in the negative direction, the electrons in the beam are slowed down. When the beam electrons are speeded up, a region of high electron density is formed; when the beam electrons are slowed down, a sparse region of low electron density is formed. The average speed of electrons in traversing from the gun end to the collector end of the tube is not changed by this bunching process. Such electrons are speeded up to the same extent they are slowed down. The physics of kinetic energy has something very definite to say regarding such a situation; no net energy is imparted to, or extracted from, the electron beam as the result of such bunching. However, the dense electron regions have stronger magnetic fields than the sparse electron regions. This means that the helix is subjected to a change in flux, which from basic electrical principles, we know must induce a voltage wave in the helix. This induced or 'reinduced' wave grows in power level during its progress along the helix. How do we account for this action?

Progressive increase in bunching density

We note in Fig. 2.26 that, commencing from the gun end of the helix, each successive region of bunched electrons is denser than that preceding it. This is because, once bunched, electrons tend to remain so during their transit down the tube. Consequently, a given bunched region is repeatedly subjected to the bunching process as it interacts with the periodic electric field of the signal wave. The bunching thereby becomes cumulative, with electron density increasing as the electrons in the beam are bunched during their transit from the gun to the collector end of the tube. In similar fashion, the sparse regions become less dense with successive interactions with the electric field of the signal wave. The natural consequence of this situation is that the electromagnetic induction of the second helix wave, the one we previously referred to as being reinduced, becomes more pronounced as we approach the collector end of the helix. In this way, the power level of the second helix wave grows from one induction cycle to another. This growing wave does extract energy from the electron beam. In giving up energy to the helix wave, the beam electrons are slowed down. In order to effect an energy transfer, it is necessary that the reinduced wave be propagated slightly slower than the average speed of beam electrons. This condition is readily brought about by adjustment of the d.c. helix voltage. (It so happens that when the signal wave and the electrons have identical propagation speeds, the reinduced wave, i.e., the growing wave, lags behind the average electron transit time.) Although basically sound, our analysis has been simplified in that we considered by *one* growing wave. This growing wave itself velocity modulates the beam to produce electron bunches, which in turn reinduces another growing wave on the helix. This process occurs numerous times so that the ultimate growing wave is a composite of many such waves.

Backward-wave oscillator

In the backward-wave oscillator, we permit reflection of the composite growing wave back to the input end of the helix. Oscillation occurs at the frequency at which the phase of returned signals most strongly reinforces the helix wave at the input end of the tube. No external cavity or other resonant device is necessary to tune the oscillations. Rather, the frequency of oscillation is governed by helix voltage. We see that travelling-wave tubes, in common with klystrons and magnetrons, effectively *use* the transit time of electrons. Conversely, in ordinary vacuum tubes, transit time of the electrons in the cathode-plate gap establishes the upper limit to high-frequency performance.

Oscillator theory in terms of the universal amplifier

The largest class of oscillators make use of an active device capable of providing power amplification, a resonant circuit and a feedback network or provision. The unique association of these functional elements identifies the oscillator and generally is responsible for its operating characteristics. Inasmuch as there are many oscillator circuits, as well as a number of active devices, it would greatly simplify our study of oscillators if it were not necessary to deal individually with the large number of permutations acruing from these facts. It forunately turns out that most oscillators can be made to work with most active devices. To a considerable extent, oscillators don't care what kind of active device is employed.

This is not to say that optimum performance will necessarily attain from a blind selection of the active device, or that one can always directly substitute one active device for another. However, with little experimentation, or calculated modification of component values, most oscillators are capable of satisfactory performance with two, three or more active devices, any of which may have been chosen for a variety of reasons. This makes it unnecessary to treat the theory of oscillation separately for different active devices and leads to an elegant conceptual approach to the subject.

Dealing with a 'universal' active device such as the amplifier enables us to focus attention on the circuitry aspects of oscillators. Once the basic principles of an oscillator circuit are grasped, the designer or experimenter is then free to consider an npn or pnp transistor, an n-channel or p-channel JFET or MOSFET, various type of op amps or an electron tube. The choice may be governed or influenced by considerations of frequency, power level, availability, cost or system compatibility. But, a single basic theory will apply, no matter what choice is made. Moreover, it should be noted that most active devices are basically three-terminal devices. Even those with additional terminals still operate as amplifiers via the interaction of three 'main' terminals or electrodes. Dual-gate MOSFETs and screen-grid tubes are examples of such cases.

Figure 2.27 shows the symbol of a universal amplifier this book will use in order to exploit the aforementioned facts. The triangle symbol has long been used in block diagrams to depict various types of amplifiers. Although it is also well known as the symbol of a particular active device–the op amp–its proper use in this connection requires the input terminals to be labelled + and −, or noninvert and invert. The technical literature generally supports

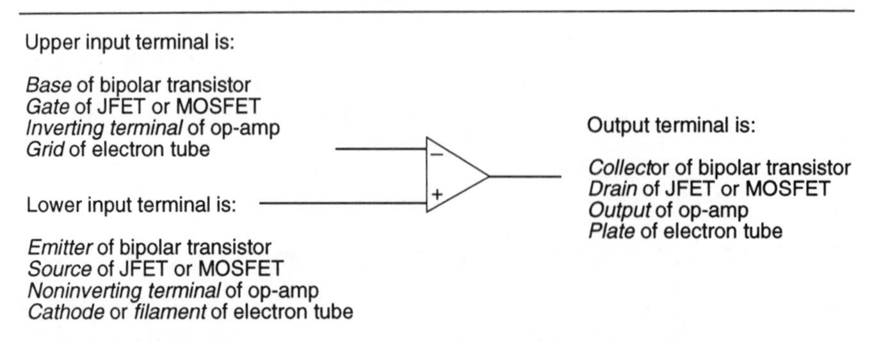

Fig. 2.27 *The universal amplifier symbol. Inasmuch as most oscillators can be made to operate with a variety of active devices, it is too restrictive to discuss the operating principles in terms of a single device. The triangle symbol is commonly used to depict all kinds of amplifiers and is, therefore, appropriate for use with oscillator circuits*

the use of the unmarked triangle as a representative amplifier. In other words, the triangle symbolizes the function of amplification. It is in this sense that it will be used to facilitate the explanation of various oscillator circuits. Later, specific symbols of various active devices will be used in practical oscillator circuits to serve as examples that serve as guides to actual applications.

A number of active devices that can be represented by the universal active device or universal amplifier are shown in Fig. 2.28. These probably embrace the vast majority of devices presently used in oscillators. One could, however, postulate others. For example, thyristors such as the SCR and the PUT could conceivably be included. Also there are new solid-state devices being developed that could assume importance in oscillator applications. An example is a hybrid device in which a power MOSFET and a power bipolar transistor are monolithically joined so that the net device has a field-effect control terminal and a pn junction output section. Also, it is possible to cite other logic devices in addition to those illustrated that can be used in oscillator circuits. The Schmitt trigger, for example, is sometimes seen in crystal oscillator circuits used for clocking logic systems.

Some considerations in the selection of semiconductor devices for oscillators

Figure 2.29 shows a Colpitts oscillator configured about an active device. The active device can be any of the symbolic representations shown. This is because they all produce power-gain and are, essentially, three-terminal devices. (Tetrode and pentode tubes, as well as dual-gate MOSFETs can be viewed as basically three-terminal amplifying devices with auxiliary control-

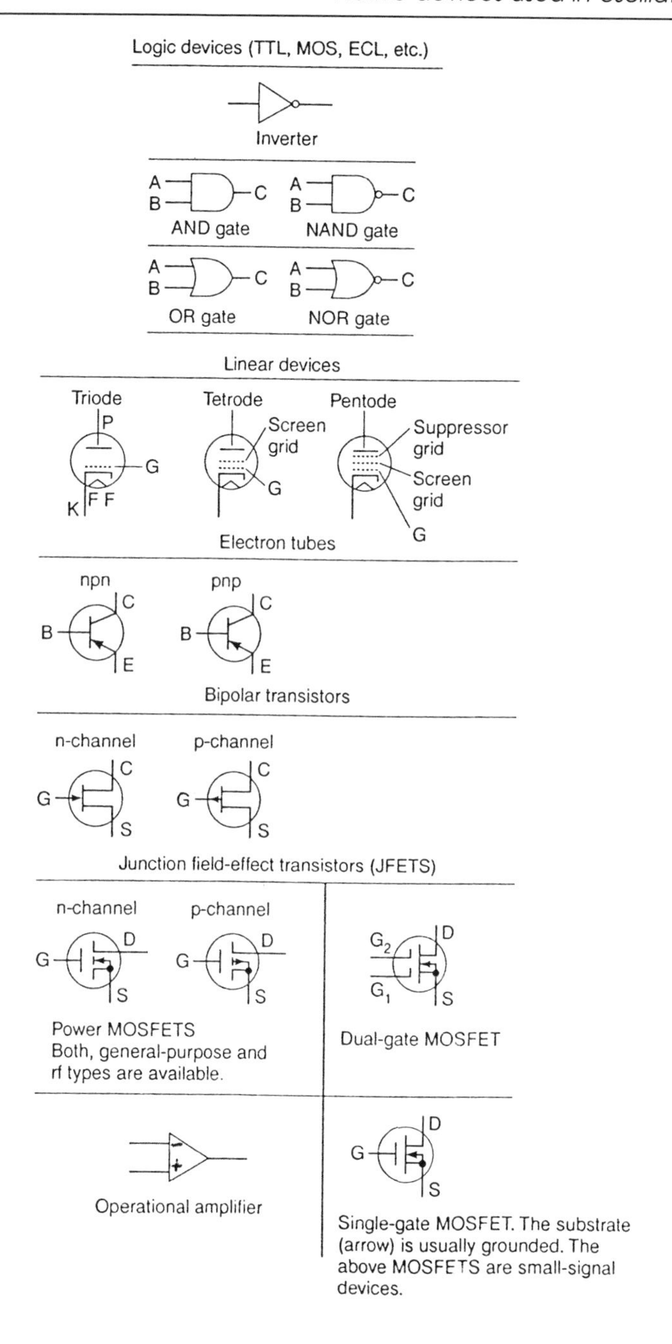

Fig. 2.28 *Active devices that can be used in oscillator circuits. The linear devices are readily usable in common oscillators such as the Colpitts of Fig. 2.29. Logic gates are also often used in various oscillator circuits*

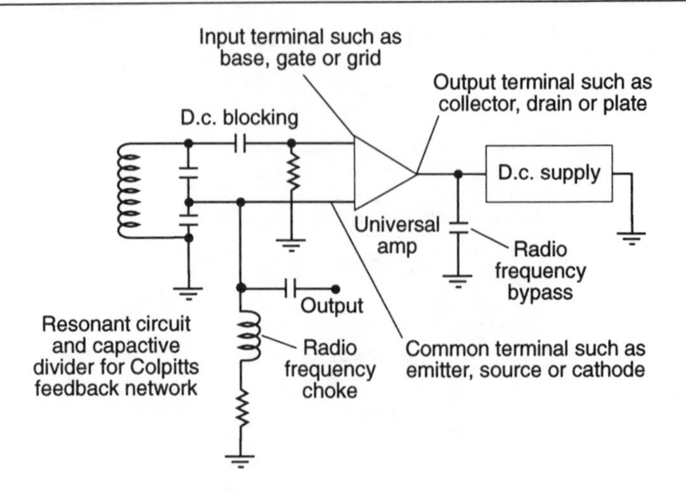

Fig. 2.29 *A basic Colpitts oscillator shown with a three-terminal amplifying device. The implication of such a schematic diagram is that a wide variety of active devices can be employed in the circuit. The triangle is, accordingly, the symbol of the so-called universal amplifier*

elements.) However, there are significant differences in performance, convenience and cost among the active devices. Some excel at high frequencies, while others would merit consideration if we were striving for high-power-levels. With some, it is easier to stabilize frequency against the effects of temperature, operating voltage or ageing. Some are more immune to physical and electrical abuse. The impedance-levels of some are less degrading to resonant circuits than are others. And, naturally, some will be found more compatible to a particular system or application than others.

Some examples will illustrate factors likely to influence the choice of an active device for an oscillator. To begin with, if it is desired to attain rapid warm-up and to conserve space and weight, the electron tube would necessarily have to give way to solid-state elements. If high input-impedance is required, the bipolar transistor would generally not be as good a choice as the MOSFET or JFET. And, in this instance, power MOSFETs are commonly available, whereas power JFETs remain the objects of overseas experimentation. Linear op amps can be configured into very precise oscillators because their high gains enables the simultaneous use of positive and negative feedback. Thus, an op amp would be a good device to use in a sine wave oscillator where the emphasis is on wave purity rather than power output.

Sometimes the selection of the active element is based upon more subtle considerations than, say, frequency or power capability. For example, the design of a solid-state dip meter (intended to perform similarly to the tube-type grid-dip meter) might not work satisfactorily with a bipolar transistor because the bias current of the transistor tends to mask the rectified oscillator current. And a MOSFET lacks an input diode needed to provide the 'grid' current. It so happens, however, that a JFET has an input diode somewhat analogous to the grid-cathode diode of the tube. The analogy is close enough so that the JFET can almost directly substitute for tube circuits in this application. But for most other oscillator purposes, it is undesirable to operate so that the input diode is driven into forward conduction.

The nearly universal use of bipolar transistors for all kinds of oscillators may not always reflect the best wisdom of choice. For example, a bipolar transistor oscillating at a frequency well down on its current-gain characteristic may cause trouble with low-frequency 'parasitics' because of its relatively high gain at low frequencies.

The JFET is a very popular device for almost direct substitution in the crystal oscillator circuits that were long used with tubes. Many crystals are intended for oscillation in the parallel-resonant mode and are stipulated for a certain frequency with the proviso that a capacitance of 32 pF be presented by the oscillator circuit. Because of the relatively-high input impedance of the JFET, this condition is readily met. Moreover, the input impedance of the JFET is less influenced by temperature or by d.c. operating-voltage than is the case with bipolar junction transistors. Thus, the frequency stability of the crystal tends to be good with very little effort expended in circuit optimization. This is particularly true because the JFET contributes a negligible amount of heat compared to the tube implementations.

A workhorse JFET crystal oscillator is shown in Fig. 2.30. With the capacitive divider and radio frequency choke designated, this circuit will serve many communications system needs in the 1–5 MHz range. Note the diode shunted from gate to ground. Such an element was not seen in the otherwise-similar tube circuits. Whereas it may be acceptable to drive the grid of the electron tube into its positive region, analogous forward-bias of the gate-source diode in the JFET is not desirable. The tube continues to amplify when there is grid current, but the JFET does not operate properly when there is gate current. Gate current causes abrupt limiting of amplification; at best it causes harmonic generation—at worst, it can stop oscillation, especially if the crystal is not too active. Besides the manufacturer's specifications do not advocate such operation. It turns out in practice that a small positive voltage at the gate is not detrimental, but means are needed to prevent heavy conduction. To this end, the 1N914 diode is used to divert most of the current that would otherwise circulate in the gate-source circuit of the FET.

The equal-valued capacitive dividers are generally satisfactory for this Colpitts circuit. (In bipolar transistor versions of this circuit, special attention is required for the capacitive divider beyond the mere compliance with feedback criteria. But, in the FET oscillator, there is much less need to swamp out or isolate the device's input impedance from the crystal.) The

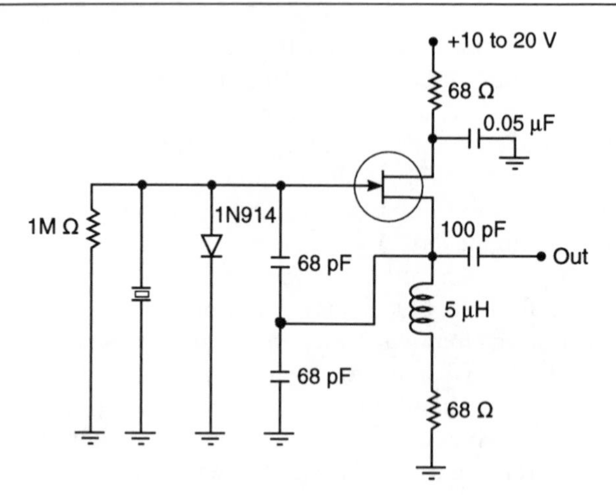

Fig. 2.30 *FET oscillator for use with parallel-resonant 1–5 MHz crystals. This is a good workhorse oscillator combining simplicity with reliable operation. Many practical oscillators using specific semiconductor devices are discussed in Chapter 5*

combination of the resistance and inductance in the source lead enables d.c. bias and high impedance from ground to be developed independently. At the same time, any stray resonances that might exist in the radio frequency choke are prevented from being troublesome by the series resistance, which acts as a Q spoiler. Although a 100 pF output coupling capacitor is shown, its size should be reduced to the minimum value consistent with the needs of the load. A small trimmer capacitor of, say 15 pF or so, can be connected across the crystal to provide a limited range of frequency variation.

The d.c. operating voltage and currents required by solid-state devices can exert considerable influence on the selection of such an element for an oscillator circuit. Although this is often merely a matter of convenience, the input and output impedances of oscillator devices depend upon *how* a given output power is produced. If low voltages and high currents are used, the impedances will be low. The converse is true for high d.c. voltages and low currents, in which case tube-like impedances may be simulated. When high frequencies and high-power levels are both required from an oscillator, the input and output impedances may tend awkwardly low and this would be aggravated by the use of a high-current low-voltage device. The trouble is that low impedances generally require low inductances and high capacitances in the resonant and impedance-matching circuits–nanohenries of inductance and tenths of microfarads of capacitance would, for example, be awkward or downright impractical to implement. Other things being equal, the power MOSFET displays higher impedance levels, at least at the input, than bipolar transistors.

Another factor in the selection of the semiconductor active device has to do with d.c. operating polarity. Whereas tubes were restricted to one polarity, bipolar transistors are available in their npn and pnp versions so that positive or negative power sources can be used. Similarly, MOSFETs are made in n-channel and p-channel fabrications. In either case, however, it must be ascertained that the device with the desired polarity is available with the other features needed for desired oscillator performance.

A factor sometimes overlooked in production runs of oscillator circuits is the wide tolerances generally pertaining to bipolar transistors. Whereas the amplification factor or transconductance of electron tubes may have been within a $\pm 10\%$ variation, the current-gain factor (β) of a given type of transistor may vary as much as three to one. An oscillator that performs well with a 'hot' transistor may be reluctant to start with a transistor having a more typical beta value.

The logic devices usually operate in their digital modes and produce square wave outputs. The linear devices perform well in the oscillator circuits formerly used with tubes and operate in a similar fashion. Not shown are other solid-state devices that are useful in relaxation or negative resistance oscillators. Such devices would include the SCR, the PUT, the unijunction transistor (UJT), the tunnel diode, the lambda diode, the gunn diode and the IMPATT diode. It can be seen that the designer, technician and hobbyist now has a greatly expanded array of devices and circuits to consider.

3 Theory of oscillators

We now examine the oscillation theory in somewhat more detail than has been expedient during our discussions of oscillating elements and oscillation-provoking devices as separate entities of the oscillator. It is appropriate that we combine these basic oscillator components and investigate the overall effect. In particular, we will concern ourselves with feedback oscillators. All feedback oscillators present a negative resistance to the resonant tank circuit. The essential difference between those oscillators we specifically designate as of the negative-resistance variety and those we call feedback oscillators, involves a practical consideration that does not appreciably affect the theory of operation. In the negative resistance oscillator, the tank circuit can be separated from the device exhibiting negative resistance. In the feed-back oscillator, the oscillation provoking device no longer displays negative resistance if we sever it from the resonant tank. This is illustrated in Figs 3.1a and b, where the tank circuits are shown disconnected and replaced by a milliammeter. At Fig. 3.1a negative resistance is still present as indicated by a decrease in output current as output voltage is increased. At Fig. 3.1b an increase in output voltage causes an increase in output current, indicating positive resistance rather than negative.

The tunnel diode

The tunnel diode is a two-terminal semiconductor device with direct capability of inducing oscillation in an associated tank circuit. From a practical viewpoint, this provides the most illuminating demonstration of the principle of negative resistance. The tunnel diode incorporates a pn junction and bears some similarity to junction rectifiers and zener diodes. In particular, its characteristics suggest a zener diode already in its avalanche condition without any externally provided bias. This being the case, a slight

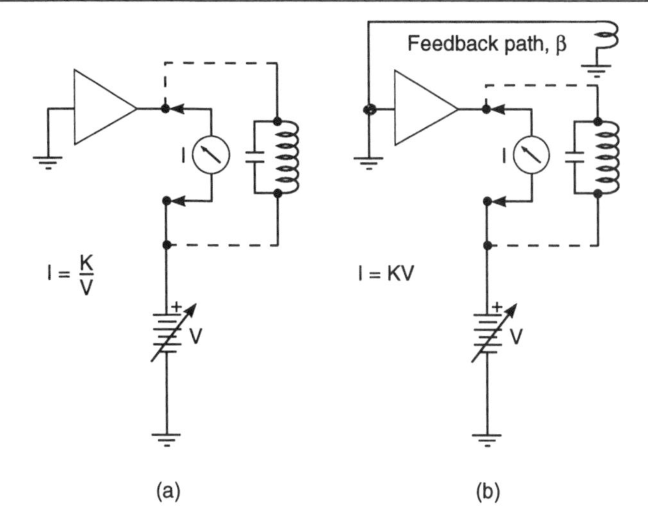

Fig. 3.1 *Comparison of negative-resistance oscillator and feedback oscillator. The negative-resistance device displays negative-resistance without external feedback. a, Negative-resistance oscillator. b, Feedback oscillator*

bias voltage supplied from an external source suffices to project the internal charge condition of the junction from zener or reverse conduction to forward conduction. In the accomplishment of this transition, an increase in bias voltage is accompanied by a decrease in bias current. Such a voltage-current relationship corresponds to negative resistance. The negative resistance thereby produced cancels a numerically equal amount of ordinary, or positive, resistance in resonant circuits. Under proper conditions, this results in sustained oscillation.

In other oscillation-provoking elements, the maximum frequency of operation is ultimately limited by transit time of whatever charge carries are involved. That is, the time required for electrons, holes or ions to traverse a distance between two electrodes establishes a limit beyond which higher frequency oscillation is not attainable even though the effects of stray capacitances and inductances are circumvented. In the tunnel diode, however, the transit time is infinitesimal with respect to generation of radio frequency energy. The high-frequency limit imposed by this factor is at least 10,000,000 MHz. Thus, we find extensive usage of the tunnel diode in microwave techniques and radar applications.

The two oscillation modes of the tunnel diode

When the tunnel diode is used in conjunction with reactive elements, two types of oscillation may be obtained. Either sinusoidal or relaxation oscilla-

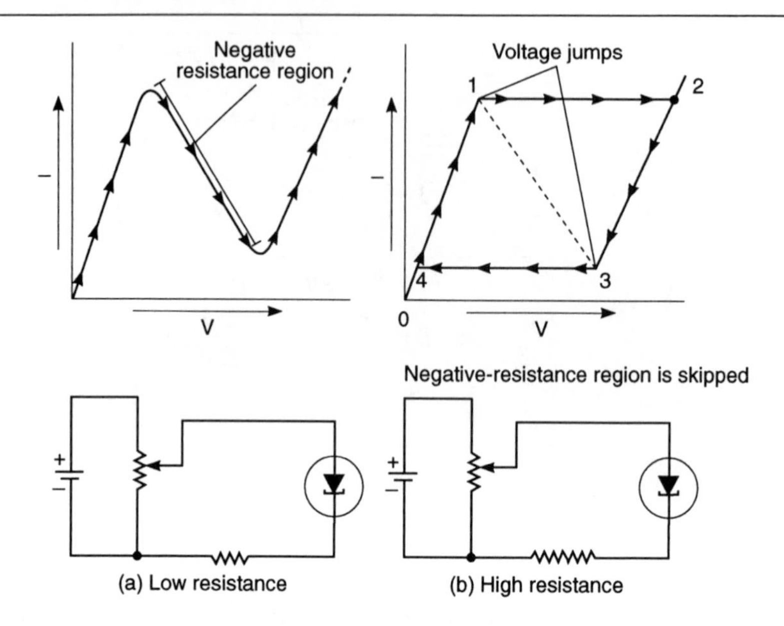

Fig. 3.2 *Operational modes of tunnel diode corresponding to two types of oscillators*

tions will occur, depending very largely upon the regulation of the d.c. power source and its voltage. This is best illustrated by the test circuits of Fig. 3.2. In Fig. 3.2a a *low* value of limiting resistance is used. As the potentiometer is advanced, the 'N' characteristic curve of the tunnel diode is followed. If the operating point were fixed in the centre of the negative-resistance region, sinusoidal oscillations could be produced by properly associating an LC tank with the circuit. This is so because the *negative* resistance of the tunnel diode could then be used effectively to cancel the dissipative effect of ordinary or *positive*, resistance in the LC tank.

In the event the limiting resistance is *high*, that is, if the d.c. regulation of the power source is relatively poor, an entirely different situation exists. This is depicted in Fig. 3.2b. As the potentiometer is advanced, the tunnel diode current increases along its characteristic from 0 to 1. Advancing the potentiometer further however, does *not* cause the operating point to trace its path along the downward slope of the negative-resistance region. Instead, an abrupt *jump* occurs with the operating point establishing itself at 2. Now, let the potentiometer be slowly returned toward its zero voltage position. The operating point will follow the characteristic downward from 2. After thus approaching point 3, the operating point will *not* locate itself on the slope of the negative-resistance region. Rather, a very slight further retardation of the potentiometer will cause an abrupt voltage *jump* from 3 to 4. If we now *advance* the potentiometer, the 1, 2, 3, 4, cycle will be repeated. Thus, we see

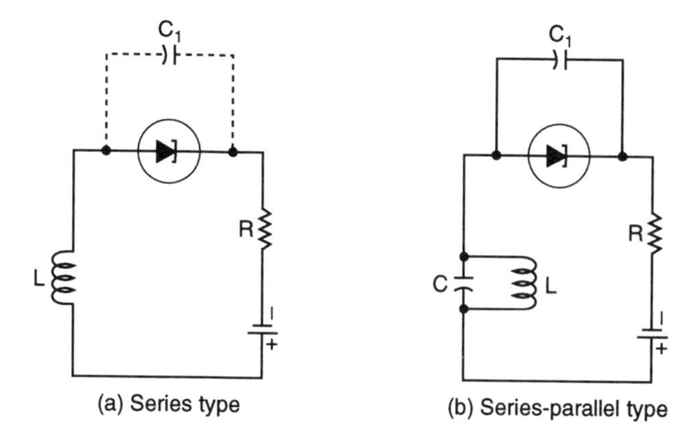

Fig. 3.3 *Tunnel-diode sine-wave oscillators*

that when the series resistance is high, *switching transitions* occur with essential avoidance of the negative resistance region. When such a circuit is properly associated with an energy storage element such as an inductor, relaxation-type oscillations can be generated.

Tunnel-diode sinusoidal oscillators

The simplest tunnel-diode oscillator for generating sinusoidal waves is shown in Fig. 3.3a. The frequency f is found by the formula:

$$f = \frac{1}{2\pi}\sqrt{\frac{1 - (R \times 1/r)}{L \times C_1}}$$

where:

f is the frequency in hertz,
R is resistance in ohms,
r is the negative resistance of the diode in ohms,
L is inductance in henrys,
C is capacitance in farads.

What this circuit lacks in schematic complexity is compensated for by the abundant, and often highly mathematical, technical literature dealing with the theory of tunnel-diode operation. For the circuit of Fig. 3.3a, a basic relationship establishes the nucleus for more extensive analysis of oscillatory action. The time constant given by the net positive resistance of the circuit in conjunction with the inductance must be equal to, or less than, the time constant due to the capacitance of the diode in conjunction with the negative resistance of the diode. Expressed symbolically, $L/R = rC_1$, where:

L is the inductance in the circuit; R is the net circuit resistance; r is the negative resistance of the diode; and C_1 is the effective capacitance of the diode.

In order to obtain approximately linear operation, that is, dynamic excursion of the operating point over the essentially straight portion of the negative-resistance region, L/R must be equal to or less than rC_1. This is most readily brought about by adjustment of R. However, a disadvantage of this circuit is that this relationship is difficult to maintain by variation of any *single* parameter. In other words, if we desire sinusoidal waveshape, we cannot readily change frequency by the convenience of varying the inductance, L.

A more stable and useful circuit is shown in Fig. 3.3b. The frequency of oscillation of this circuit is found by the formula:

$$f = \frac{1}{2\pi}\sqrt{\frac{1}{L\ (C + C_1)} - \frac{(1/r)^2}{C_1\ (C + C_1)}}$$

where:

f is the frequency in hertz,
r is the negative resistance of the diode in ohms,
L is inductance in henrys,
C and C_1 are capacitances in farads.

Here, the criterion of oscillation is described by a somewhat more complex relationship:

$$R = \frac{1/r}{(2\pi f)^2 + (1/r)^2}$$

where:

f is the frequency of oscillation.

In this circuit, the frequency can be readily tuned without causing the dynamic operating range to exceed the approximately linear segment of the negative-resistance region. In both circuits, R cannot *exceed* a certain value without causing a change in oscillatory mode. When R is too large, relaxation type oscillations are produced due to abrupt switching between operating points. Although R is depicted as a single series resistance, in practical oscillators R is generally the *net* resistance due to a voltage divider and the d.c. resistance of the inductance, L.

Push–pull tunnel-diode oscillator

For the purpose of approximately doubling power output, we may devise push–pull versions of the basic single-diode oscillators. The circuit of Fig. 3.4 is a particularly useful one. The tunnel diodes can be biased to operate in

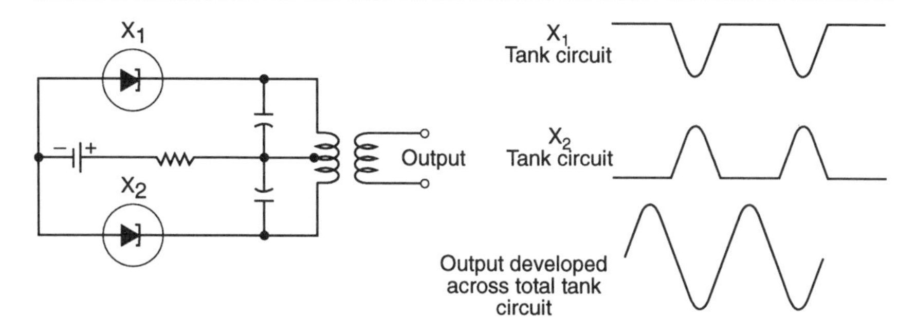

Fig. 3.4 *Push-pull tunnel-diode oscillator and waveforms*

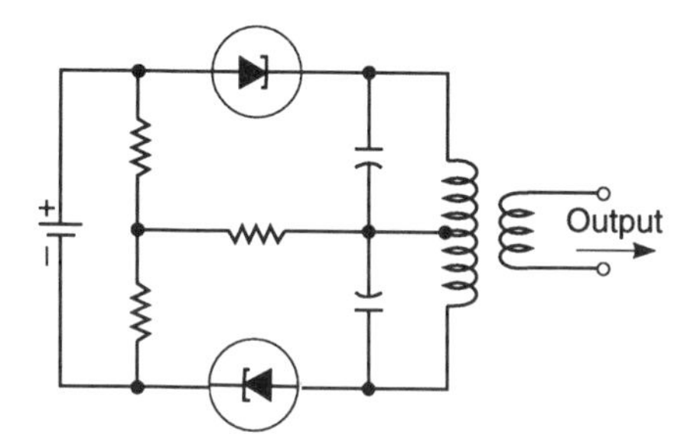

Fig. 3.5 *Cascade tunnel-diode oscillator*

the relaxation or switching mode. This tends to provide relatively high output and reliable oscillatory action. At the same time, the output can be a fairly good approach to a sinusoid. This is brought about by the combined effects of harmonic discrimination by the tuned tank, and the tendency of the push–pull configuration to cancel even-order harmonics.

Cascade tunnel-diode oscillator

An alternate methods of increasing output power is to connect simple tunnel-diode oscillators in series, at the same time modifying the d.c. bias network so that only one battery or other d.c. source is needed. This technique is not limited to two diodes; rather, any number can be so cascaded. A cascade arrangement involving two diodes is shown in Fig. 3.5 for comparison with the push–pull circuit of Fig. 3.4. It should be noted that the relative polarity of the diodes differs in the two circuits.

With unfortunately chosen parameters, the circuit of Fig. 3.5 can operate so that the phasing in the two tank sections does not induce additive emfs in the output winding. This operational mode can be discouraged by the use of relative high ratios of C to L in the tank sections.

Whereas, in the push–pull circuit, the two diodes alternate switching

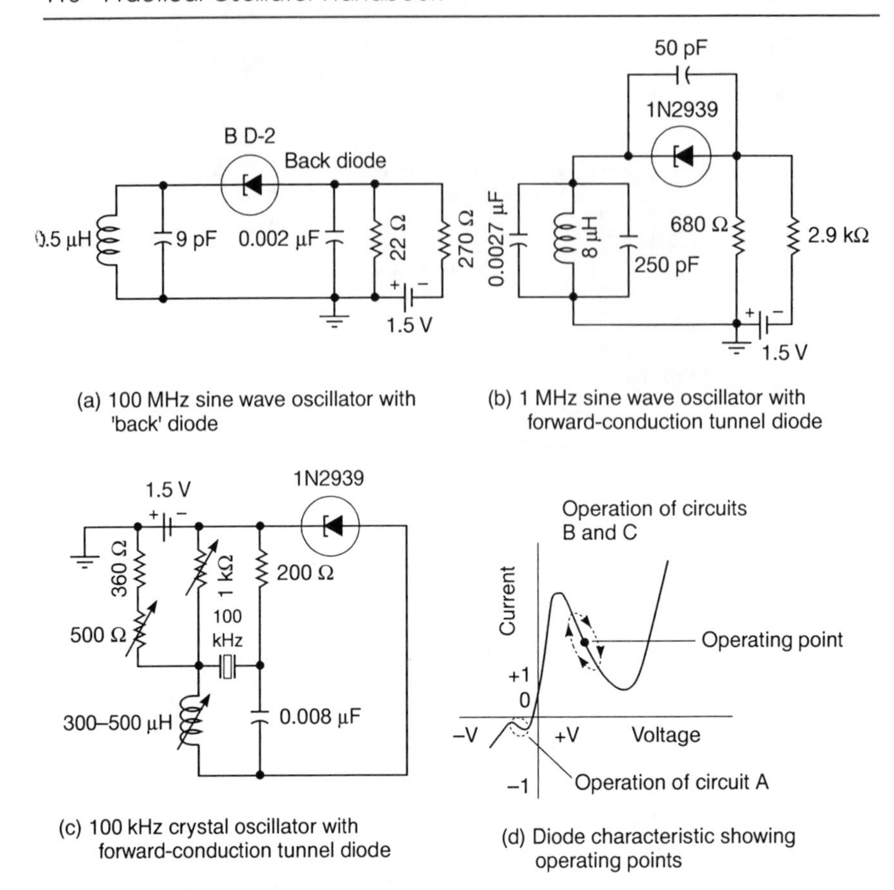

Fig. 3.6 *Tunnel-diode oscillators*

states, in the cascade arrangement both diodes undergo their switching cycles simultaneously. That is, the switching cycles of the transistors are in phase. Biasing for linear mode operation is also feasible in this arrangement.

Several practical oscillators are shown in Fig. 3.6. In Fig. 3.6a, the circuit performs as the series-parallel type. Because of the high frequency, the physical capacitor that ordinarily is shunted across the diode is not required. In this case, the internal capacitance of the diode serves the function of C_1. The circuit in Fig. 3.6b is of the series-parallel type as shown in Fig. 3.3b. It will be observed that all three circuits illustrated in Fig. 3.6 make use of voltage-divider networks rather than the single series resistance of the basic circuits of Fig. 3.3. This is because most practical power sources, certainly commercially available batteries, provide voltages far above the several hundred millivolts needed to bias tunnel diodes in the middle of their negative-resistance regions. Also, the circuit of Fig. 3.6a utilizes the so-called 'back diode'. This is a tunnel diode that operates under *reverse*, rather than forward, bias conditions. These diodes generally operate at low power levels than forward-conduction types, but at the same time permit the use of inductors with readily obtainable values. It should be appreciated that R in the basic circuits of Fig. 3.3 is, in the general case, the *net d.c.* resistance due to voltage-divider networks, coil resistance, and internal resistance of the power source.

The crystal-controlled oscillator of Fig. 3.6c is unique, being suggestive of tube or transistor oscillators employing a crystal in the feedback loop rather than directly as in a conventional tank circuit. In this arrangement, oscillation would not exist without the crystal because the net a.c. resistance R would then exceed the criterion for sinusoidal oscillation. At the same time, the voltage is sufficiently divided down so that relaxation oscillation cannot take place. With the crystal in the circuit, the two resistances are effectively in parallel for the frequency corresponding to series resonance of the crystal. Thus, the net value of R is lowered, enabling oscillation to be sustained in the LC tank circuit.

The Class-C feedback oscillator

Our emphasis on feedback theory will cast much light on the operating principles of the frequently encountered oscillators in tube technology, the Class-C oscillator. Although the Class-C oscillator is, at least in terms of one viewpoint, a switching circuit that shock-excites an LC tank with narrow pulses, we will find that much insight can be gained by treating such an oscillator as an overdriven feedback oscillator. Thus we will, in many instances, consider the Class-C oscillator as a feedback oscillator in which operation extends above and below the linear portion of the transfer characteristics of the tube. Strongly excited oscillators using solid-state devices tend to operate in Class-B rather than Class-C. It remains true, however, that their resonant tank circuits are also shock-excited by pulse waveforms. (Calling them Class-C oscillators is a hangover from tube practice, but generally does no harm.)

In previous discussion, oscillators in general were described as being devices that automatically converted d.c. power to a.c. We are now confining our interest to feedback oscillators in particular, a feedback oscillator being an amplifier that supplies its own input signal. It does this via a feedback loop. The feedback loop is a signal path that permits a portion of the output voltage (or current) to be reimpressed at the input of the amplifier. Generally, but not necessarily, the feedback loop is a circuit external to the amplifier proper. Simply returning a portion of the amplified output signal to the input of the amplifier is not in itself sufficient for oscillation; the phase, or polarity, of the returned signal must be the same as

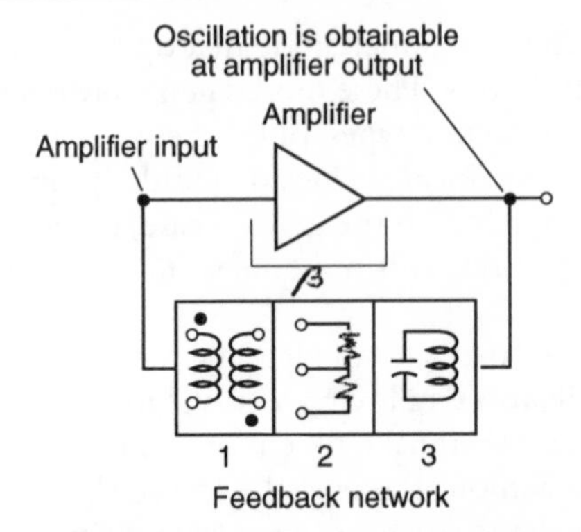

Fig. 3.7 *Basic elements of the feedback oscillator*

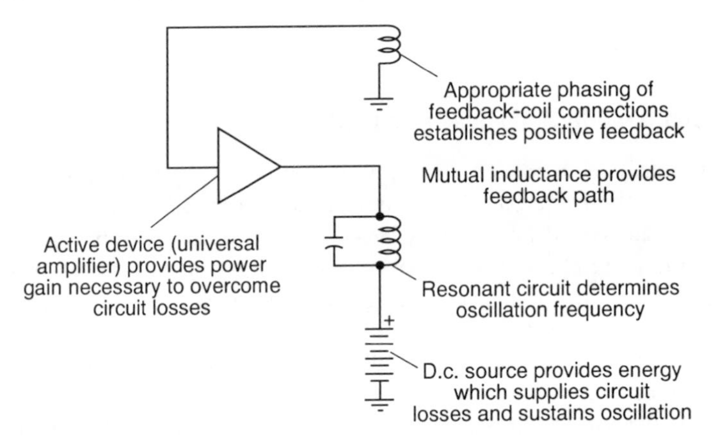

Fig. 3.8 *Function of the elements in a feedback oscillator*

the input signal responsible for its appearance. In this way, the original signal is reinforced in amplitude. The reinforcement is cumulative; a given change in the input signal produces a change in the amplified output signal, which in turn assists the input signal in continuing the process. The theoretical build-up to infinite amplitude inferred by this process is, in practice, limited to a finite value. We have here a condition of regeneration, and the feedback is said to be positive. Oscillation occurs at a frequency determined by a resonant tank, which functionally becomes part of the feedback network. The components of the feedback network in feedback oscillators are shown in Figs 3.7 and 3.8.

The question of original signal voltage

Before defining additional criteria for the onset of oscillation in the generalized system, it is expedient that we resolve a possible conflict in describing oscillator operation. Technical authors generally assume an oscillator is already in oscillation, then proceed to relate the manner in which such oscillation is sustained. In so doing, the reader is sometimes left to wonder how the oscillations began. It is easy to see that an oscillator is an amplifier that supplies its own input signal, but we are tempted to ask, from whence came the output signal when the oscillator was first turned on. Surely, there could be no output signal without an input signal, and conversely no input signal without an output signal.

Initiation of oscillation build-up

A simple two-terminal element, such as a length of wire or a resistance, is not, strictly speaking, passive. Thermally excited electrons produce minute, but for some purposes significant, voltages across the ends of such elements. Specifically, we have a *random frequency generator* that develops an effective rms voltage, V_N, defined by the equation;

$$V_N = \sqrt{V_0^2 + V_1^2 + V_2^2 + V_3^2 + V_\infty^2}$$

where we designate all frequencies from d.c., V_0, to infinite frequency, V_∞. At any instant, any frequency has a statistically equal chance of occurring. That is, V_N is pure white noise.

In real life, a definite resistance associated with a finite bandwidth is involved, and so is a temperature in excess of absolute zero. Thus, it is found the V_N can be calculated as follows:

$$V_N = \sqrt{4k \times T \times BW \times R}$$

where:

V_N is the rms value of the noise voltage comprising all frequencies within bandwidth BW,
BW is in hertz,
T is the absolute temperature (centrigrade plus 273°),
R is the resistance of the two-terminal element in ohms,
k is Boltzmann's constant (1.38×10^{-23}).
At room temperature of 21 °C, this equation simplifies to:

$$V_N = \sqrt{1.623 \times 10^{-20} \times BW \times R}$$

The presence of V_N at the input of a positive-feedback amplifier constitutes the 'start' signal for initiation of a regnerative build-up process culminating in sustained oscillation at the frequency dictated by the frequency-determining elements contained in the feedback loop. Thus, we see that any practical amplifier or oscillator always has a 'built-up' input signal.

The switching of power-supply voltage often induces transients in the resonant tanks that shock-excite oscillations, thereby providing the initial signal for cumulative build-up. In some cases, it would be difficult to ascertain whether power-supply transients or thermal-noise voltage was the more influential in starting the regenerative cycle leading to amplitude oscillation. In any event, a definite time is required for the build-up process. The higher the Q of the resonant tank, the greater is the build-up time. Build-up time also depends upon feedback, being shorter with heavy feedback. A crystal oscillator with just sufficient feedback for oscillation can consume several seconds in attaining constant level operation. Such an oscillator obviously would not be suitable for interruption by a telegraph key.

Effect of fixed bias on spontaneous oscillation build-up (tube circuits)

A tube oscillator may be provided with a source of fixed bias in addition to grid-leak bias. Such an arrangement is often desirable for large tubes as a protection against cessation of oscillation. We know that grid-leak bias exists only during oscillation. If for some reason such as excessive loading, oscillation should stop, a ruinous plate current could quickly destroy the tube, unless some other biasing scheme was incorporated to provide negative grid voltage in the absence of oscillation. However, if a fixed bias is too high, neither thermal noise voltage nor circuit transients would suffice to start the build-up process. Such a situation would, of course, exist if the fixed bias projected the tube into its cut-off region. The tube would provide no amplification under this condition. Sometimes an oscillator with such a bias arrangement is encountered in diathermy or high-frequency heating applications. Oscillation is started by momentarily shorting (through a relatively low resistor) the fixed-bias supply.

Effect of positive feedback on gain of an amplifier

Because of the popularity of high-fidelity audio equipment, it is common knowledge that negative feedback imposes a reduction of amplifier gain, but permits certain desirable improvements such as extended frequency response, better stability and relative independence from active-device and circuit variations. Some oscillators use negative-feedback loops, and we shall have more to say regarding negative feedback later. However, our primary concern in oscillators is the nature of positive feedback. Let us designate the voltage gain of an amplifier without feedback by K, and its voltage gain with feedback by K'. We wish to establish a relationship between K and K'. We intuitively suspect that the amount of feedback returned from the amplifier output to the amplifier input will be involved. Let us then designate the fraction of the output voltage reapplied at the input by the symbol β, which is plus for positive feedback (regeneration) and, appropriately, negative, for inverse feedback (degeneration). These quantities will then be related according to the equation:

$$K' = \frac{K}{1 - K\beta}$$

We see from this equation that positive feedback increases the gain of an amplifier.

Physical interpretation of infinite gain

Further contemplation of the equation reveals that it is possible for the denominator to approach or even become zero. Such a condition means the gain of our amplifier becomes infinite. What, indeed, is the practical significance of infinite gain? Actually, before infinite gain is ever attained, the amplifier bursts into oscillation at a frequency that produces an in-phase relationship between input and feedback signals. The in-phase relationship, in turn, is governed by the sum of all reactive components (inductors and capacitors) in the amplifier circuit. We see why a regenerative detector, being essentially an amplifier with controllable positive feedback, permits such high amplification from a single amplifier. In Fig. 3.9 we see a simple amplifier circuit with several modifications added to make the relationship

$$K' = \frac{K}{1 - K\beta}$$

more meaningful with respect to our study of oscillators. In this formula, K is the gain without feedback, and K' is the gain with feedback. We see that β, the portion of output voltage fed back to the input, is determined by a voltage divider. The transformer is connected to invert the phase of the amplified signal appearing in the circuit. Familiarization with this basic circuit will provide a great deal of insight into the operating principles of feedback oscillators, notwithstanding the complexity of the feedback network or the novelty of various approaches in other circuitry sections of the oscillator.

An interesting feature of feedback oscillators is shown in Fig. 3.10. Here we see a graph depicting the minimum value of β required for oscillation as a

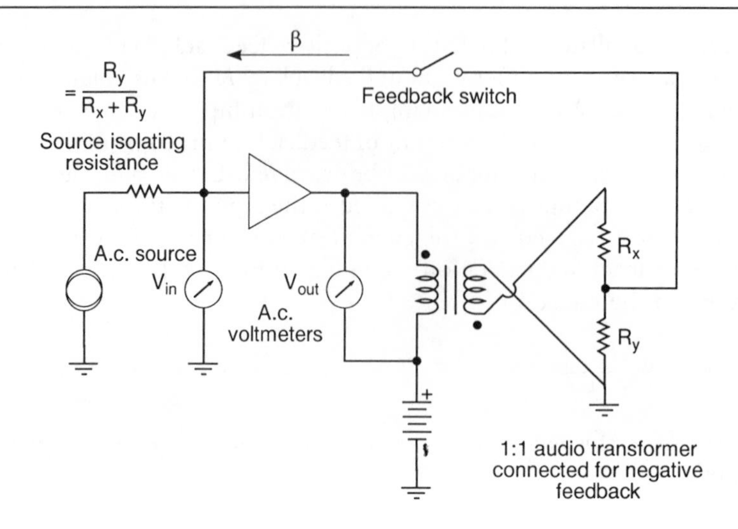

Fig. 3.9 *Test circuit for study of feedback. Feedback equations can be confirmed with the input and output a.c. voltages measured in this negative-feedback amplifier*

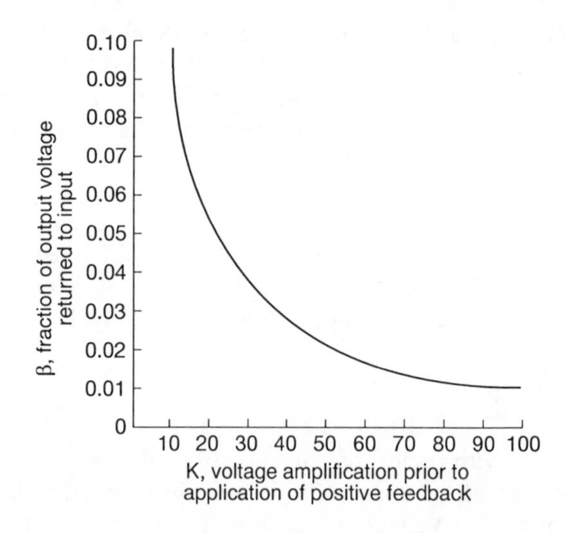

Fig. 3.10 *Minimum feedback signal required for oscillation*

function of K, the nonfeedback voltage gain of the amplifier. Significantly, the graph illustrates that, when K is high, β is relatively low. Usually, it is desirable to have K high from the standpoint of frequency stability. This is particularly true in tube circuits when the frequency-determining tank is in the grid circuit, for then the plate and load circuits will be better isolated from the grid circuit than would be the case with low K and the necessarily high β.

Feedback and negative resistance from the 'viewpoint' of the resonant tank

We have seen that oscillation in a feedback oscillator occurs when the product $K\beta$ is equal to one. This gives us the mathematical condition of infinite amplification in the equation relating amplification with feedback, K', to amplification without feedback, K. We should now reflect upon the physical significance of equating $K\beta$ to unity. This condition signifies that we have apportioned just sufficient output voltage to compensate exactly the feedback network losses. Generally, the main component of the feedback network is the resonant tank. Consequently, the tank has applied to it a signal ($K\beta$) of sufficient amplitude to compensate for the damping action of the losses, which would otherwise cause oscillations to die out in the event they were somehow started. We see that the basic philosophy involving action on the resonant tank is not qualitatively different than that occurring in the negative-resistance oscillator. In both the negative-resistance oscillator and the feedback oscillator, the tank circuit becomes oscillatory when its equivalent resistance is matched by a numerically equal negative resistance, the negative resistance being a generator or amplifier condition that cancels the effect of ordinary losses, or 'positive' resistance. The equivalent resistance, R_0, of a parallel-tuned tank is $L/R_S C$. In the dynatron oscillator then, the dynamic plate resistance of the tube, $-r_p$, must be equal to the quantity $L/R_S C$ for oscillation to take place. It is likewise true that the resonant tank of the feedback oscillator 'sees' $L/R_S C\ \Omega$ of negative resistance when feedback conditions are just right to sustain oscillation (this relationship is pictured in the diagram of Fig. 3.11).

Basic consideration of phase of feedback signal

It has been mentioned, but without emphasis on phase conditions, that the product of $K\beta$ must be at least unity for the condition of oscillation to be attained in the feedback oscillator. It is, however, also necessary that the fractional output signal, β, returned to the input be in-phase with the input signal.

An amplifier of conventional circuit configuration (grounded cathode, emitter, source or 'invert' terminal) causes a phase displacement of 180° between output and input signals. A 180° phase shift corresponds to polarity inversions; when the input signal is undergoing a positive excursion in its cycle, the amplified output signal is at a corresponding negative point in its cycle. We see that direct feedback from output to input must produce

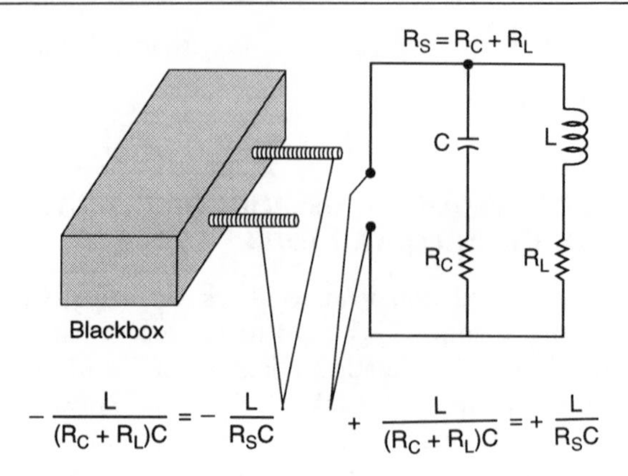

Fig. 3.11 *The mathematical principle of oscillation reconciles the differences between feedback and negative-resistance techniques*

negative, not positive, feedback; that is, β is negative. When we insert a negative value of β in the expression $K/(1 - K\beta)$, the resulting amplification given by this expression shows a decrease, not the necessary increase towards the mathematical identity of infinite gain. We see the need for a means of shifting the phase of the output signal 180° before returning it to the input. Beta would then be positive. The way in which this is accomplished in practice largely accounts for the many different circuits of feedback oscillators. We know some of the ways wherein such a phase shift can be obtained. Transformers, autotransformers, series-resonant tanks, additional amplifier stages and RC networks can be used for this purpose. (If the oscillator is configured around a grounded grid, base or gate amplifier, no phase shifting is required in the feedback network to bring about the condition of positive feedback necessary to produce oscillation. The same is true of voltage-follower amplifier circuits.)

Feedback under various phase conditions

We have discussed the condition of positive feedback, and have pointed out the opposite nature of negative feedback. What about the phase conditions that correspond neither to ideal in-phase or out-of-phase signal return? It is only by investigating this more general condition that we can gain a comprehensive insight into the basic concept of oscillation in the feedback amplifier. This purpose is best served by studying feedback conditions in an amplifier with a feedback loop over a wide frequency range. We may even suppose that the amplifier is of the so-called 'negative feedback' type in which positive feedback is not deliberately designed into the circuit. By measurement or computation of the amplitude and relative frequency range,

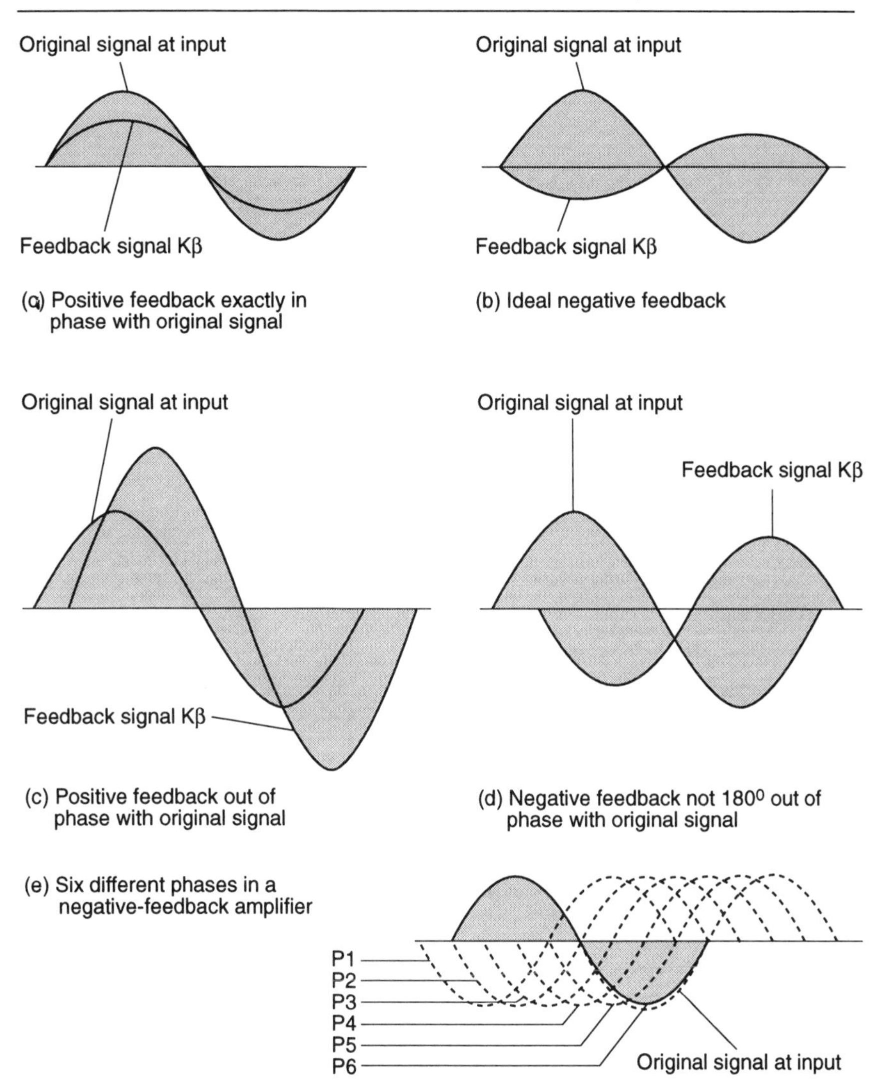

Fig. 3.12 *Various feedback conditions in amplifiers and oscillators*

it will always be found that negative feedback departs more and more from the ideal out-of-phase relationships as frequencies much lower or much higher than the design bandwidth are checked. Indeed, at certain low and high frequencies, it will be found that the feedback becomes positive. This is a natural consequence of the total circuit reactances. Actual capacitors and inductors tend to bring about this condition at low frequencies, whereas at high frequencies stray capacitance and transformer leakage inductance are more generally involved. If, at a frequency corresponding to exact in-phase positive feed-back, the loop gain $K\beta$ equals or exceeds unity, the amplifier becomes oscillatory at that frequency.

The foregoing principles are illustrated in Fig. 3.12. Positive feedback is pictured at Fig. 3.12a, with the feedback signal exactly in phase with the original signal. Oscillation will result if the amplitude of the feedback signal is sufficient, that is, if $K\beta = 1$. At Fig. 3.12b, an ideal negative-feedback condition is shown, with the feedback signal exactly 180° out of phase with the original signal. At Fig. 3.12c there is positive feedback not exactly in phase with the original signal. Oscillation cannot occur even for large amplitudes of the feedback signal. Figure 3.12d shows negative feedback not exactly 180° out of phase and Fig. 3.12e shows six different phases of feedback. At phase P6 there is an exact in-phase relationship, and oscillation will take place if $K\beta = 1$.

The practical obstacle to infinite build-up

One of the basic operating features of all oscillators is the slowing down and ultimate cessation of the oscillatory build-up rate. We intuitively know that this happens, for otherwise the oscillatory amplitude would theoretically increase forever; tremendous circulating currents would exist even in a physically small tank circuit, and the process would come to a halt only as a result of a burnt-out inductor or capacitor. There is nothing in the nature of the resonant tank that otherwise tends to restrict the amplitude build-up of oscillations when the effect of its losses are neutralized by feedback or negative resistance. Therefore, we must investigate the oscillation-provoking device. In so doing, we find that all such devices, no matter how divergent their operating principles, possess one common feature. As circuit elements they all show nonlinear relationships between voltage and current when voltage or current is sufficiently small or large relative to the values within the centre of more ordinary operating regions. For example, the Class-A amplifier, which we almost instinctively associate with linear or proportional signal transfer, is linear only when a.c. input and output voltages are small enough to limit operation over straight-line portions of the graphs representing input and output voltage-current relationships. Let us consider several specific instances in which nonlinearity exerts a braking effect upon oscillation build-up.

Unique amplitude limiting in Class-C tube oscillators

Because tubes are depletion devices (normally 'on') and because the grid can draw current in normal operation if hard-driven, an amplitude limiting phenomenon takes place that is not emulated in solid-state devices.

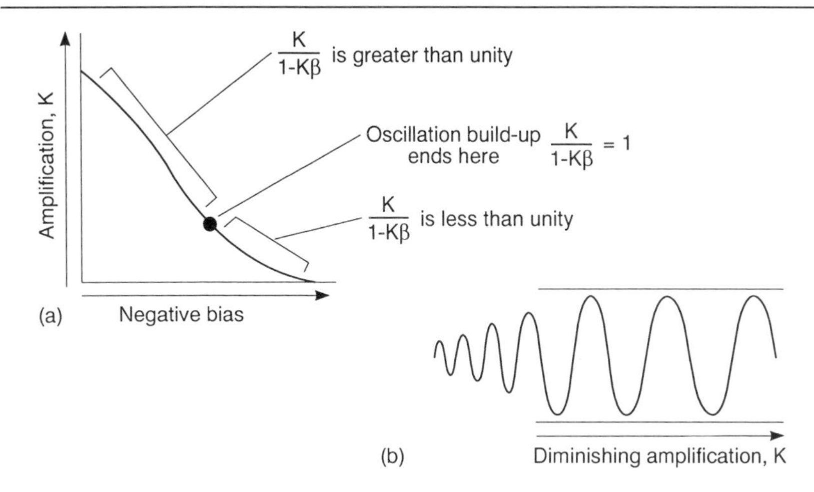

Fig. 3.13 *The amplitude-limiting action of automatic bias in an oscillator. a, Amplification as a function of the negative input bias. b, Amplitude limiting caused by oscillation-produced bias. This type of limiting can occur in tube circuits, but not in solid-state oscillators*

In the Class-C circuit it is most common to employ automatic bias. The input of the amplifier is biased towards cut-off by the oscillation itself. Before such an oscillator commences operation, no negative input bias is developed and signal amplification, K, is at, or near its maximum value. Assuming that the value of $K/(1 - K\beta)$ is, or exceeds, unity, oscillation build-up will start. As drive builds up, the grid will draw current through the automatic bias resistance ('gridleak') and develop a voltage drop across this resistance. As a consequence of the rectifying action of the positively-pulsed grid and the filtering action of the input capacitor, a negative input bias is developed across the capacitor. The greater this bias, the greater is the reduction in the amplification. This is shown in Fig. 3.13a. It is seen that sufficient negative bias will bring the amplifier to the point where no further build-up of oscillation amplitude can take place.

From Fig. 3.13b it can be seen that oscillation amplitude initially builds up rapidly, but soon its rate of increase slows down. The increase comes to a halt when sufficient negative bias is developed so that $K/(1 - K\beta)$ is unity. Thus, the Class-C oscillator, because of oscillation dependent bias, has a built-in amplitude-limiting mechanism. Although this amplitude limiting scheme initially appears applicable to the JFET, also a depletion device, such is not the case. It is not permissible to drive the input diode of the JFET into its current consuming region.

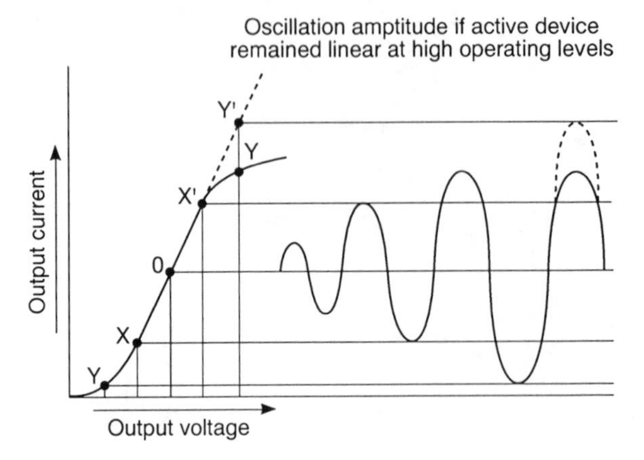

Fig. 3.14 *Limiting action in a Class-A oscillator. Because all active devices depart from linearity at low and high operating levels, equal increments of voltage produce equal increments of current over a restricted amplitude range*

Amplitude limiting in Class-A oscillators

In the Class-A oscillator, we do not have the automatic amplitude-limiting action of d.c. input bias. Here, the large-signal nonlinearity in the output voltage-current relationship exerts the limiting function. A tube operating in a Class-A oscillator is a voltage controlled device; a grid-voltage change produces a plate-current change, which by its flow through an impedance inserted in the plate circuit develops a voltage change of greater magnitude than the original grid signal. We see that although a voltage is fed back to the grid in an oscillator, the voltage is the result of plate current. Consequently, the ability of the tube to produce plate-current changes in response to grid-voltage changes is directly relevant to feedback oscillator operation.

Suppose oscillation is building up and has attained an amplitude corresponding to operation beyond that straight-line relationship between output voltage and current (see Fig. 3.14). With further build-up, a given increment of input no longer produces as large output–current change, as when the characteristics remained linear. Consequently, less voltage drop is developed across the tank, which, in turn, implies a smaller increment of feedback signal delivered to the input. Thus, the rate of build-up has slowed down. The slowdown continues as operation is extended further into the nonlinear regions. Ultimately, a point is reached wherein further build-up no longer pays its way, so to speak.

The foregoing points are illustrated in Fig. 3.14. An increase in output voltage, such as might occur over the region XY produces a smaller increase in output current than does a similar variation of output voltage within the XX′ region. This can cause the value of $K/(1 - K\beta)$ to fall below unity, thereby preventing the oscillation amplitude to exceed approximately the YY level.

Springs, weights and oscillating charges

If one were not already knowledgable about the matter, the oscillatory nature of the simple LC 'tank' circuit might well be anticipated from certain clues. One often cited is the analogy with a mechanical system comprising a ceiling-supported spring with an attached weight. In such an arrangement, a momentary force which alters the rest state of the spring initiates a periodic up–down motion which ultimately dies down because of various energy-dissipative frictional forces. (Otherwise, the physical oscillation would persist indefinitely.) It should be noted that the distorted spring stores *potential* energy, this being the energy 'on tap' due to the position or the static state of an entity.

In contrast, the weight, or more precisely, the mass represented by the weight, is invested with *kinetic* energy or the energy of motion. Mass also manifests the property of inertia, which is the tendency to persist in the type and magnitude of motion occurring at any instant. Significantly, once provoked, the alluded energy manifestations then reinforce one another so that there is a periodic interchange of these energies in the spring-mass system. During such oscillation, the inertial behaviour of the mass *opposes* the force received from the spring and reverses its direction. Accordingly, the energy residing in the oscillating system is continuously changing from 'pure' potential energy to 'pure' kinetic energy. Of course, most of the time, the system energy is partitioned between the two types.

In the electrical system, the charged capacitor represents a stressed situation with the potential energy revealed by the voltage developed across the capacitor plates. This potential energy is stored in the electrostatic field and is on tap by forcing current to flow into an associated circuit or load. When teamed with an inductor, this electrostatic energy is depleted as the current then builds up in the inductor. The presence of current in the inductor implies storage of the received energy in the magnetic field of the inductor. Note that the charged capacitor was capable of storing *stationary* charges as reservoir of potential energy. In contrast, the kinetic energy stored in the inductor stems from *moving* charges. From Lenz's law, we know that a changing current in an inductance develops a counter EMF so polarized as to *oppose* the voltage responsible for the current change. This, then, gives rise for a cooperative effort on the part of the capacitor and the inductor to *reverse* periodically the direction of current in the system.

The current reversals take place twice each cycle and represent interchange between potential and kinetic energy, as in the mechanical system. Ideally, the net energy of the oscillating system would remain constant and oscillation would go on indefinitely. Because of the dissipative losses in any practical LC circuit, sustained oscillation requires energy replenishment, repeated 'shocks' or negative resistance.

Another clue to the oscillatory nature of the simple LC circuit could occur to the mathematically-oriented mind. Such perception is often advantageous in that it can lead to quantitative as well as qualitative information. For the sake of simplicity, let's again deal with the simple LC circuit in ideal form assuming zero dissipative losses. In such a set-up, Kirchhoff's laws tell us that the algebraic sum of the voltages at any instant must be zero. We often see this expressed as

$$L\frac{di}{dt} + \frac{q}{C} = 0$$

where:

L represents the inductance,
di/dt represents current change with respect to time,
q represents the charge stored in the capacitor,
C represents capacitance.

For our purpose, we need not deal with units, but since $L\frac{di}{dt}$ is the counter EMF or *voltage* generated by the inductance, it will be convenient to also express the capacitor charge as an equivalent voltage. This is easily done via the following identity:

$$\frac{q}{C} = \frac{1}{C}\int i \, dt$$

Re-writing our Kirchhoff equation, we then have

$$L\frac{di}{dt} + \frac{1}{C}\int i \, dt = 0$$

Next, this equation is differentiated with respect to time in order to make it a true differential equation:

$$L\frac{d^2i}{dt^2} + \frac{i}{C} = 0$$

which is more conveniently dealt with in the form,

$$\frac{d^2i}{dt^2} = -\frac{i}{LC}$$

Such a differential equation is packed with information because its solution(s) reveal much about the circuit behaviour. Solutions of differential equations require skill, practice and often experimental procedures involving 'educated guesses'. Thus, it can be confirmed that the following identity will 'work' as a solution to this differential equation:

$$i = a(\cos \omega_o t) + b(\sin \omega_o t)$$

where:

a and b are arbitrary constants determined by the starting condition of the assumed oscillation, ω_o represents the resonant frequency of the oscillation in radians per unit time.

Without going into details or solving for numerical values in a specific circuit, two significant facts have now been revealed about the general nature of the ideal LC circuit:

The sine–cosine relationship that is found to fit into the differential equation suggests the oscillatory nature of the current. It is moreover found that if we let $\omega_o = 1/\sqrt{LC}$, the differential equation is unambiguously solved. We then interpret ω_o as the 'natural' or resonant frequency in radians per unit time. The resonant frequency in hertz is then $\omega_o/2\pi$, assuming L is expressed in henries and C in farads.

Interestingly, other mathematical strategies also 'work' in the differential equation of Kirchhoff's law of the LC circuit. Significantly, all of these solutions ultimately reveal sinusoidal oscillation and culminate in the same expression for the resonant frequency. For example, those accustomed to representing a.c. circuit parameters by exponential notation will find that the following expression for current also works as a solution to the differential equations:

$$i = A\mathrm{e}^{j\omega t} + B\mathrm{e}^{-j\omega t}$$

where:

A and B are arbitrary constants relating to the starting conditions of the assumed oscillations.

Although the mathematical steps will necessarily differ from those used before, exactly the same harmonic motion of the charges in the circuit will be revealed, and exactly the same relationship between the LC parameters and the natural or resonant oscillation frequency will be attained.

In many practical LC circuits, the combined energy dissipation from all causes can be represented by a series resistance between the L and C elements. In such a case, the application of the above procedures commences with the Kirchhoff equation written as

$$L\frac{\mathrm{d}i}{\mathrm{d}t} + Ri + \frac{1}{C}\int i \, \mathrm{d}t = 0$$

This leads to the resonant frequency,

$$\omega_o = \sqrt{\frac{1}{LC} - \frac{R^2}{4L^2}}$$

where again $f_o = \frac{\omega_o}{2\pi}$ in hertz. In most electronic applications, the Q of the LC circuit is sufficiently high (R is sufficiently low) so that the simpler and more familiar relationship $f_o = \frac{1}{2\pi\sqrt{LC}}$ is entirely satisfactory.

Amplitude limiting in negative-resistance oscillators

In the dynatron negative-resistance oscillator, it might appear that unrestricted oscillatory build-up would take place because the amplifying function of the tube is not used. Here, we might postulate that a decrease in gain is not meaningful inasmuch as the tube behaves directly as a negative resistance. Although such an argument is sound with respect to the scope of its implication, it falls short of accounting for the overall principle of operation. In much the same way that the would-be inventor of a perpetual motion machine always finds success blocked by the manifestation of friction in some form, just so is our dynatron oscillator destined to limit oscillatory build-up via nonlinearity. The nonlinearity involved is more apparent than in the feedback oscillator, for it appears in the characteristic curve. The same is true of the transitron oscillator, the point-contact transistor negative-resistance oscillator, and other nonfeeding oscillators which supply negative resistance to the resonant tank circuit. Inspection of the negative-resistance region of the current versus voltage graph of a negative-resistance device shows that a linear, or near linear, region of negative resistance exists over a reverse slope region such as XX in Fig. 3.15. When oscillatory amplitude builds up a little bit beyond the level corresponding to operation within region XX, positive resistance is encountered, and the build-up rate is slowed down. Equilibrium is soon attained, for the positive resistance brings the circuit to its criterion of oscillation. Further build-up, with its accompanying heavier contribution of positive resistance, cannot occur because the circuit would then no longer satisfy the criterion of oscillation. The criterion of oscillation is $-r_p = L/R_SC$, where $-r_p$ is the negative resistance of the device and L/R_SC is the equivalent positive resistance of the resonant, parallel-tuned LC circuit. It is interesting to contemplate that the same criterion holds true in a feedback oscillator, that is, at the same time $K/(1 - K\beta) = 1$, $-r_p = L/R_SC$. (In this case, $-r_p$ is interpreted as the negative resistance 'seen' by the resonant tank.)

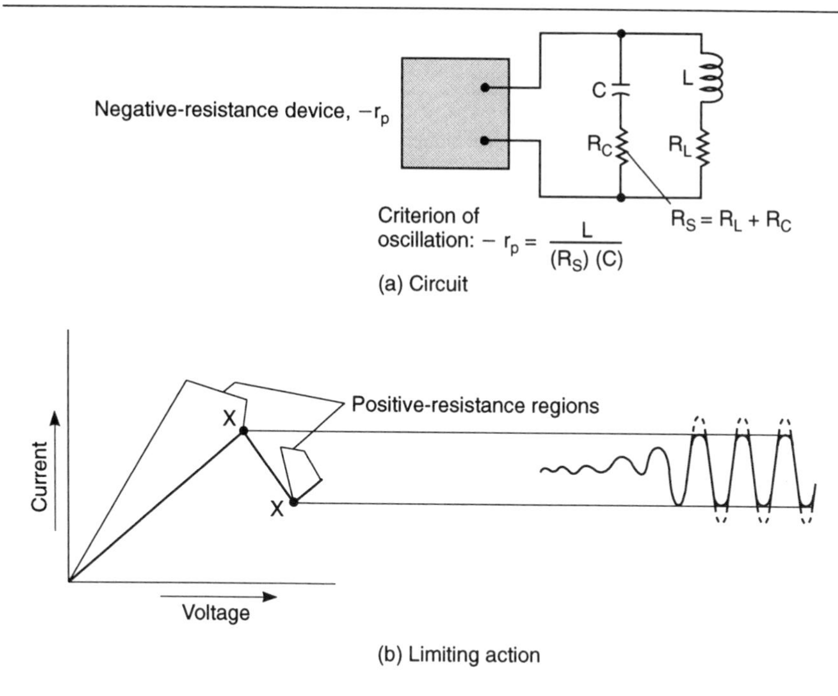

Fig. 3.15 *Limiting in a negative-resistance oscillator*

Divergent effects of bias in feedback and negative-resistance oscillators

Although an appreciation of features common to both feedback and negative-resistance oscillators cannot fail to enrich our comprehension of the oscillatory process, it would tend to defeat our purpose if we overlooked significant differences between these basic types of oscillators. We have already commented upon the fact that severance of the tank circuit destroys the negative-resistance property in the feedback oscillator, whereas negative resistance is displayed in a negative-resistance oscillator independently of the presence or characteristics of the tank. An interesting operating difference is found with regard to the behaviour of the two oscillator types. For simplicity, we will compare the electron-tube Class-A feedback oscillator with the dynatron negative-resistance oscillator. In the first circuit, increasing the value of negative bias makes oscillation *less likely*. We know this is so because the gain (K) of the tube is thereby reduced. In the dynatron, however, oscillation with a given tank circuit may not take place until the grid bias is made sufficiently negative. How do we account for this contradiction in behaviour? For a given plate voltage, more negative grid bias produces decreased plate current. This is equivalent to saying that the plate resistance of the tube has been made higher; this statement applies to both static and dynamic plate resistance. Not only is the plate resistance higher in the positive regions of the plate characteristics, but more important; it is also higher in the negative-resistance region. Thus, the ability of the dynatron to 'reach' the equivalent positive resistance of a resonant tank is enhanced as grid bias is made higher. Increased negative grid bias causes $-r_p$ to be numerically higher. Too low a negative grid bias causes $-r_p$ to be numerically less than L/R_SC, and oscillation cannot occur. The same reasoning applies to the transitron if we interpret $-r_p$ as representing the negative resistance of the dynamic screen-grid characteristic. Granted that these tube circuits are technologically passé, their behaviour remains uniquely educational.

The multivibrator

The multivibrator (Fig. 3.16) is a feedback oscillator for generating relation-type oscillations, that is, pulses or nonsinusoidal waves. Basically, the multivibrator consists of a two-stage resistance capacitance-coupled amplifier with a feedback path between output and input. The theory of operation is best approached by arbitrarily considering one stage as an ordinary amplifier and the other as a phase inverter even though their circuitries are exactly the same. Assume, also, that both stages are grossly overdriven so that they operate in both their saturated, and their cut-off regions.

Suppose a positive pulse from thermally-generated noise appears at the input of 'amplifier' A_1. The momentary increase in output current in A_1 then couples a negative-going pulse to the input of 'phase-inverter' A_2. The amplified, but inverted pulse appearing in the output circuit of A_2 is fed back to the input circuit of A_1, the polarity (positive) being proper to reinforce regeneratively the original noise transient. A rapid and cumulative build-up now takes place in the circuit until abruptly stopped by cut-off in the output current of A_2. When A_2 is driven to cut-off, A_1 is driven to a state of heavy conduction. This circuitry condition persists until the charge accumulated in capacitor C_2 can leak off through R_3.

When the voltage across C_2 is sufficiently depleted, stage A_2 again conducts; in so doing, it produces a negative-going pulse at the input of stage A_1. Now, A_1 is cut-off and remains so until the accumulated charge in capacitor C_1 leaks off through R_1. When this happens, stage A_1 reverts to its conductive state, thereby delivering a negative pulse to A_1, cutting off conduction in this stage. It is seen that the two stages are alternately on and off, the rate being determined by RC time constants. Each switching action is regenerative; once initiated in one state, the alternate stage helps drive it to completion. Such regenerative action produces steep-sided rectangular waves.

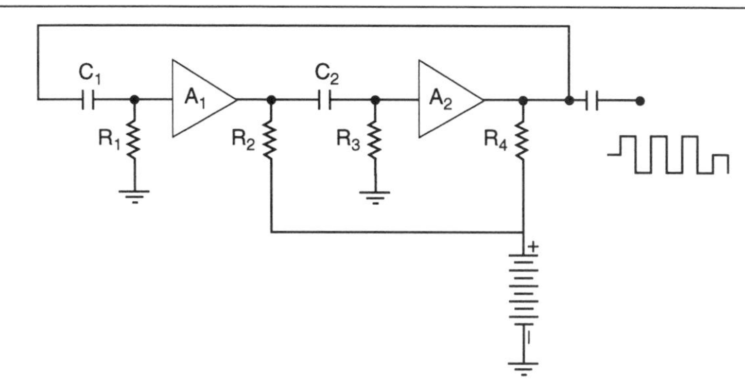

Fig. 3.16a *The multivibrator relaxation oscillator. The pulse repetition rate is approximately* $\frac{1}{R_1C_1 + R_3C_2}$ Hz.

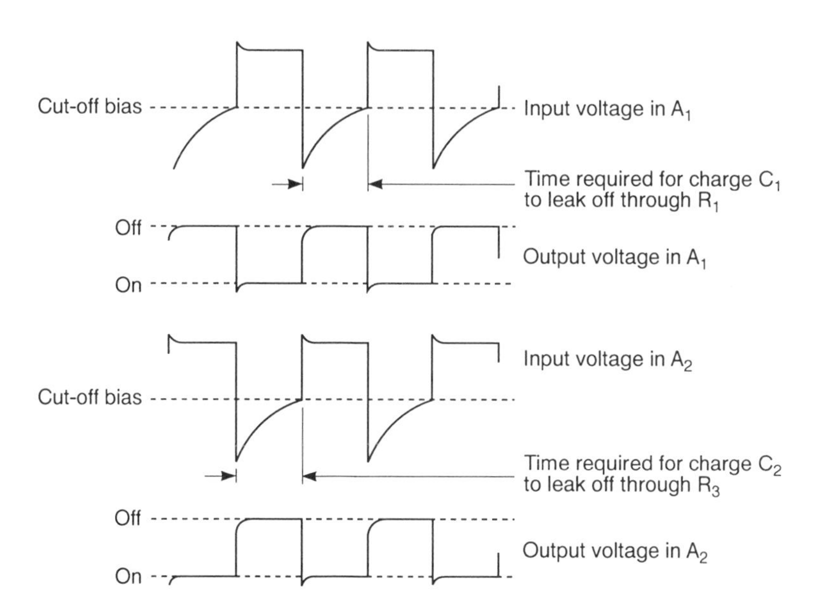

Fig. 3.16b *Waveforms in multivibrator oscillator*

The blocking oscillator

The blocking oscillator, like the multivibrator, produces relaxation-type oscillations, which have a pulse-repetition rate governed by the time-constant of an RC network. Unlike the multivibrator, the blocking oscillator contains an LC tank circuit. It is seen in Fig. 3.17 that the tank circuit

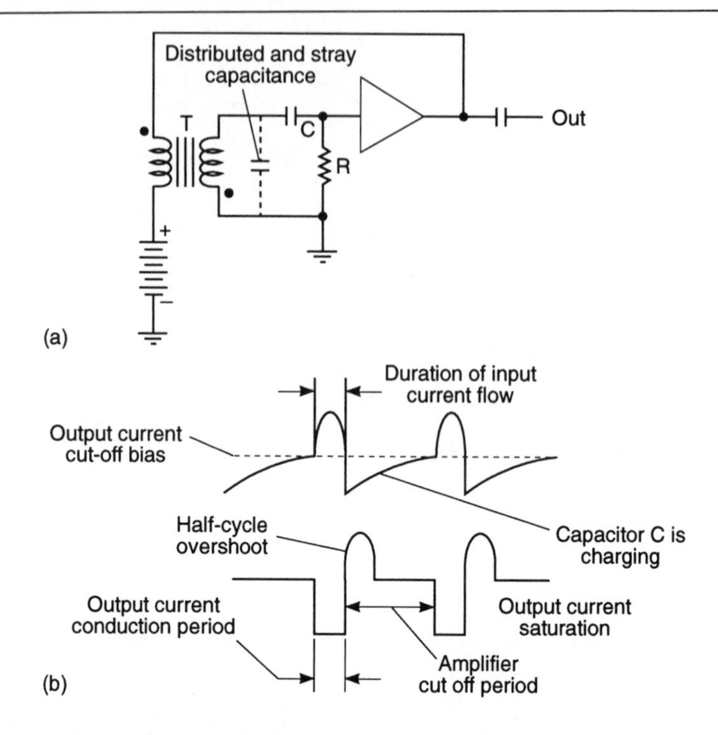

Fig. 3.17 *Basic operation of the blocking oscillator. a, Circuit. b, Waveforms*

consists of the effective inductance of the input winding of transformer T, in conjunction with distributed and stray capacitance. The schematic configuration of the oscillator is essentially that of a simple tickler-feedback circuit. However, the two windings of transformer T are very closely coupled and the time constant of the input capacitor C and resistance R is much larger than in an oscillator intended for continuous oscillation at, or near, the resonant frequency of the tank circuit; that is, C is a larger capacitor and/or R is a higher resistance than would be employed in sinusoidal feedback oscillators. Let us investigate how these differences alter the operation of the oscillator.

Action in blocking oscillator

Assume that the build-up in an oscillation cycle has begun. Output current is increasing and due to the phasing of transformer T, the input circuit of the amplifier is being rapidly driven positive, that is, into conduction. The two actions are regenerative, with one reinforcing the other. These actions proceed until output current saturation abruptly halts further increase. When this occurs, the positive drive to the input of the amplifier can maintain itself only for a relatively brief duration determined by the energy storage of the transformer. This is so because a transformer secondary develops induced voltage only when the primary current is changing in value. When the positive voltage at the amplifier-input end of the transformer finally collapses, the amplifier input is left with a high negative bias.

The development of this bias was previously discussed in the discussion on Class-C oscillators and amplifiers. Briefly, the conduction of the forward-biased (positive) input of the amplifier charges capacitor C so that its amplifier-connected plate is negative with respect of its transformer-connected plate. During the charging process, the trapped charge is overcome and cannot manifest itself as negative bias. However, shortly after the amplifier is driven into saturation, the charging current disappears, leaving the amplifier input biased with a sufficiently negative voltage to cut off the amplifier's output current. That is, a rapid transition has been made from the conductive to the nonconductive state of the amplifier. The amplifier is maintained in its *off* state until the charge trapped in capacitor C has had time to sufficiently deplete itself through resistance R. With a large capacitor and a high resistance, the quiescent time of the blocking oscillator may be many times the duration of the generated pulses.

The sudden cessation of output current causes the generation of a half-cycle of overshoot in the output winding of the transformer. The blocking oscillator is essentially a self-switching circuit that combines the operating principles of the inductively-coupled (tickler) feedback oscillator and the multivibrator. The pulse duration is determined by the natural oscillation frequency of the effective tank circuit. On the other hand, the pulse repetition rate is determined by the RC time constant of capacitor C and resistance R. It is important to keep in mind that charge time of capacitor C is very much less than discharge time. This is because the charging resistance, composed of the transformer secondary and the forward resistance of the amplifier input, is much lower than the discharge resistance provided by resistance R.

The squegging oscillator

The squegging oscillator is similar in principle to the blocking oscillator. However, instead of a single pulse being generated, discrete wavetrains of high-frequency oscillation are produced at periodically-recurrent intervals (see Fig. 3.18b). Any oscillator circuit operating in Class-C becomes a squegging oscillator when input capacitor C and/or resistance R is made so high that the charge produced in C does not have time to decay appreciably between high-frequency oscillatory cycles. The amplifier, under such conditions oscillates in spurts.

There is little difference between this type of oscillator and the blocking

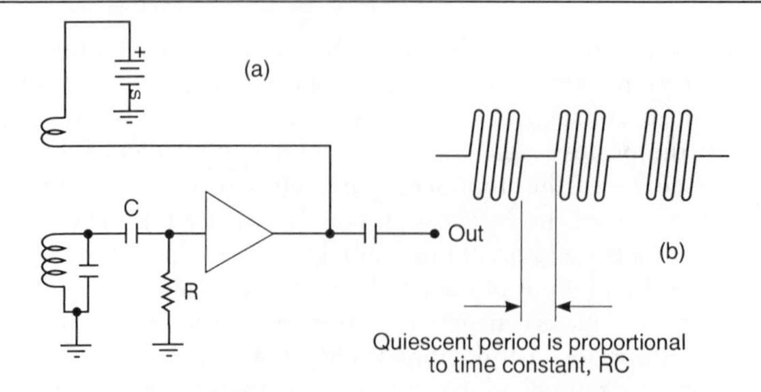

Fig. 3.18 *The squegging oscillator. a, Circuit. b, Waveform*

oscillator. The blocking oscillator generally employs an iron or ferrite-core transformer, resulting in very strong feedback. In addition, the effective Q of the equivalent tank is relatively low. This design approach results in quick disablement of amplifier operation. Conversely, in the squegging oscillator, conditions are such that many cycles of high-frequency oscillation are required to attain amplifier cut-off.

Sine wave oscillation in the phase-shift oscillator

The first network we will consider consists of three-cascaded high-pass networks. Now, at first glance, such a composite circuit may not appear significantly different than the two-similar RC networks used in the multivibrator, assuming that the separation of the networks by an active-device is of no consequence in the comparison. The important difference between the two- and the three-stage RC networks is that a 180° phase-shift is readily obtainable from the three-stage network but not from the two-stage network. This being the case, we see that such a three-stage network can be used for phase inversion in returning a signal (of correct phase for positive feedback) from the output to the input of a *single* active-device. The circuit of such a 'phase-shift' oscillator is shown in Fig. 3.19. If additionally, the active device gain is sufficient to satisfy the criterion of oscillation, $K/(1 - K\beta = 1)$, oscillation must occur.

Inasmuch as no resonant LC-circuit is present, what will be the frequency of oscillation? The oscillation frequency will be that at which the three-stage RC network produces a 180° shift in phase. Unfortunately, the attenuation of such a network is very high, that is, the factor β in $K/(1 - K\beta)$ is low. This implies the need for an active-device with appreciable gain. It can be mathematically shown that transistors must develop a current gain of 29. FETs, op amps and tubes must develop a voltage gain of 29. Because of circuit losses, such as those due to the low-impedance input of transistors,

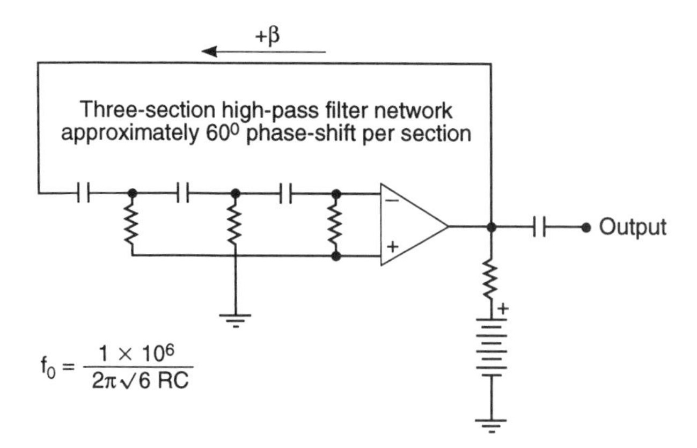

Fig. 3.19 *The phase-shift oscillator. Unlike multivibrator-circuits employing RC networks, the phase-shift oscillator delivers a pure sine wave. In order for oscillation to take place, the voltage-gain of the amplifier must be at least 29*

gains well in excess of 29 are often required in practical oscillators.

Good wave purity is attainable from phase-shift oscillators because harmonics do not produce a 180° phase-shift, and therefore cannot be supported in the oscillatory process. Phase-shift oscillators find greatest use at audio frequencies. With large, low-leakage capacitors, they are able to generate good sine waves at very low frequencies. When a transistor is used, a good rule-of-thumb is to use 10 kΩ resistors for all three shunt-arms of the RC network and for the collector load as well. Although certain frequencies and capacitor availabilities may not allow this, it will usually be found to be a good starting point.

The parallel-T oscillator

The parallel-T RC network produces a 180° phase shift at a 'resonant' frequency and has found considerable application in audiofrequency oscillators which produce good sine waves (see Fig. 3.20). Its behaviour is somewhat suggestive of a series-resonant LC circuit or of a wave trap because its resonance is accompanied by maximum attenuation of signal transfer. However, its use in oscillator circuits generally requires a certain modification. For although it can produce a 180° phase shift, this shift occurs from −90° to 90° rather than the zero to 180° needed for ordinary oscillators.

What is commonly done is to make the parallel resistance arm about one tenth the value of the series resistance arms. This brings about the needed

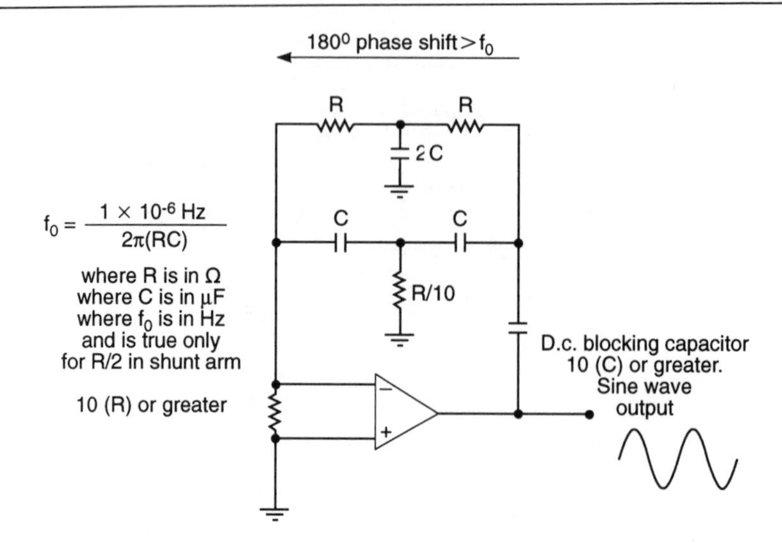

Fig. 3.20 *The parallel-T oscillator. Also known as the twin-T oscillator, this circuit is much used in low and audiofrequency applications. It is capable of exceptionally good wave purity. Note that oscillation occurs at a* higher *frequency than* f_o

type of 180° phase shift, but at a somewhat higher frequency than one would calculate for the 'natural' T network where the parallel resistance arm is a half the value of the series resistance arms. Despite the fairly drastic modification of the network, harmonics of the oscillation frequency are not accorded a favourable phase-gain relationship so a good sine wave is generated.

Providing the active device has high gain, this can be a sure-fire oscillator. Its best use is realized if the output is buffered as the load does not inadvertantly become part of the frequency determining network. Also, it is often desirable to use two transistors in the oscillator circuit proper—the first as a conventional common-emitter stage and the second as an emitter-follower. Op amps will be found even easier to implement than this oscillator.

Although vernier changes in frequency can be brought about by changing any single network element slightly about its design value, an extensive tuning-range requires the ganging together of either the three-resistances or the three-capacitances in the network.

Op-amps tend to be the best active-devices for implementing this type of oscillator. Transistor designs are most likely to be successful if *two* transistors are used—a first-stage as a conventional common-emitter circuit and a second-stage as an emitter-follower circuit. The external feedback path is then made from the output of the emitter-follower to the input of the T-network.

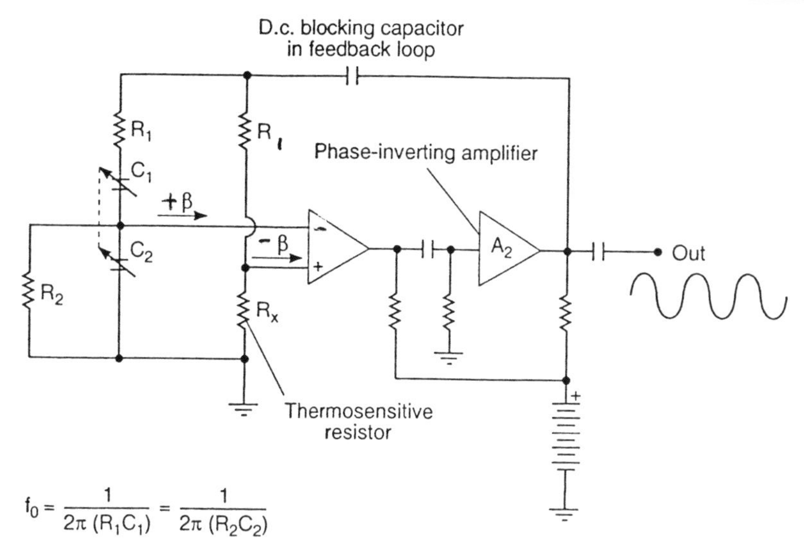

$$f_0 = \frac{1}{2\pi\,(R_1C_1)} = \frac{1}{2\pi\,(R_2C_2)}$$

Fig. 3.21 *Basic Wien bridge oscillator. When properly designed, excellent sine waves are produced by this circuit*

The Wien bridge oscillator

This is probably the most widely exploited type of RC feedback oscillator for sine waves. The name is somewhat of a misnomer, for the elements enclosed in the positive feedback loop do not make up a complete Wien bridge. The 'half-bridge' (R_1C_1, R_2C_2) involved in the positive feedback loop displays radically different characteristics than are associated with a complete Wien bridge. The response curves for a full bridge and a half bridge were previously shown in Chapter 1. Whereas proper use of the full bridge yields a nulling network or frequency rejection filter, the half bridge in the positive feedback loop of the Wien bridge oscillator produces a maximum response at the 'resonant' frequency. Furthermore, the half bridge gives us zero, rather than the 180° phase-shift associated with resonance in the complete Wien bridge. In the circuit of Fig. 3.21, the remaining two arms of the bridge are used in a negative feedback loop and are not required for oscillation. Their function in the performance of the oscillator will be discussed later.

Inasmuch as the half bridge in the positive feedback loop does not invert the phase of the fed-back signal, an additional amplifying stage is provided for this purpose. It is seen that each of the two amplifying stages produces a 180° of phase shift. The total phase shift is therefore the needed 360°. This in-phase relationship exists for only the one frequency determined by the resistances and capacitances in the half-bridge arm of the positive feedback loop. At other frequencies, this network produces a phase shift other than 360° or zero. It can be appreciated that this leads to sine wave generation—loop phase shift is not right for support of harmonics. Nonetheless, more has to be added to this oscillator for high quality sine waves.

The negative feedback loop is provided so that operation can be limited to the linear regions of the amplifiers. This leads to excellent wave purity. Resistance R_k is generally a tungsten filament lamp, or other thermosensitive element with a positive coefficient of resistance (resistance increases with temperature, and therefore with current). Because of its circuit position, this resistance provides self-bias to stage A_1. At the same time, this resistance is the shunt arm of a voltage divider consisting of itself in conjunction with series arm R_f. This network acts as a variable voltage divider in the negative feedback loop of the oscillator. The feedback is negative because the output from stage A_2 is returned to the noninvert input of stage A_1.

Suppose that the oscillation amplitude is building up. The alternating current through R_k is increasing as a result. In turn, the resistance of R_k is increasing, which causes a much higher proportion of negative feedback to be developed. A point is ultimately reached wherein the net positive feedback in the circuit is just sufficient to maintain the amplitude level. A higher build-up in amplitude cannot occur because the proportionally much greater increment of negative feedback would reduce the amplification below that needed to overcome circuit losses and maintain oscillation.

If, on the other hand, the oscillator output tends to decrease for any reason, the proportionally much greater increment of negative feedback removed from the circuit by the cooling of the lamp would permit sufficient increase in net positive feedback to very nearly restore the amplitude to its previous level.

The thermal inertia of the lamp is great enough to prevent cyclic heating and cooling above, say 100 Hz. At lower frequencies some distortion is produced due to the fact that the lamp resistance begins to follow the a.c. wave. Tuning is accomplished by ganging two elements of the half bridge in the positive feedback loop, usually the two capacitors. The frequency is inversely proportional to the capacitance, rather than to the square-root of capacitance as in LC oscillator. Consequently, a given change in tuning capacitance enables coverage of a much greater frequency range than in LC oscillators.

Loading of oscillators

The loading of an oscillator is a very important consideration. When power is extracted from a resonant tank by means of inductive or capacitive coupling to the load, a reduction in Q always occurs. This is because the load resistance is 'reflected' into the resonant LC circuit as an equivalent resistance which increases the effective value of R_S. In the interest of frequency stability, it is desirable to maintain a high operating Q of the tank. This, in itself, imposes the desirability to limit the amount of loading to a value *less* than that corresponding to maximum power in the load. There are other factors which also make such a compromise worthy of consideration. The oscillator, being a self-excited device, needs some minimum grid excitation to maintain oscillation. When we load an oscillator, the resultant reduction tank-circuit Q makes it more difficult to generate sustained oscillation. Unfortunately, the extraction of power from the tank also deprives the grid of excitation. This, in conjunction with the lowering of the Q, will cause oscillation to stop if loading is carried too far. Even before oscillation is stopped, the operation becomes unstable, with increased harmonic generation and very high plate current. Unless economic or other factors do not permit, it is always preferable to employ a lightly loaded oscillator in conjunction with a tuned power amplifier. With such an arrangement, maximum power may be delivered to the load by the amplifier with relatively little reaction on the oscillator.

Detuning effect due to loading, the Faraday shield

Another difficulty encountered with the loading of an oscillator is the detuning effect of reactance in the load circuit. Such reactance is coupled into the resonant LC circuit in a similar manner to the resistive component of the load. At radio frequencies, it is practically impossible to escape this effect. In some instances, it is convenient to retune the tank circuit to return the oscillator to its original frequency; in other cases such a procedure would constitute an annoyance at the very least. A load-coupling coil should always be located near the 'cold', that is, the a.c. ground, end of the output resonant circuit; the 'hot' end of the LC tank is susceptible to capacitive tuning by the proximity of the metal in the coupling coil. The Faraday shield, shown in Fig. 3.22, represents an additional technique to reduce this effect. A Faraday shield consists of a grid structure of parallel wires connected by a single contacting wire that is grounded. It is inserted between the load-coupling coil and the output resonant tank. Of course, this cannot isolate capacitance in the actual load circuit. A Faraday shield must not contain any closed loops, inasmuch as it is intended to act upon electrostatic lines of force but must be 'transparent' for electromagnetic induction. The transference of power from the oscillator tank to the load, by means of a capacitor, is another loading method. The detuning effect is generally more pronounced than in the case of inductive coupling. This can often be considerably overcome by using a large capacitor tapped down on the inductor of the output bank circuit, rather than a small capacitor directly connected to the 'hot' end of the tank. Inasmuch as capacitive reactance becomes smaller with

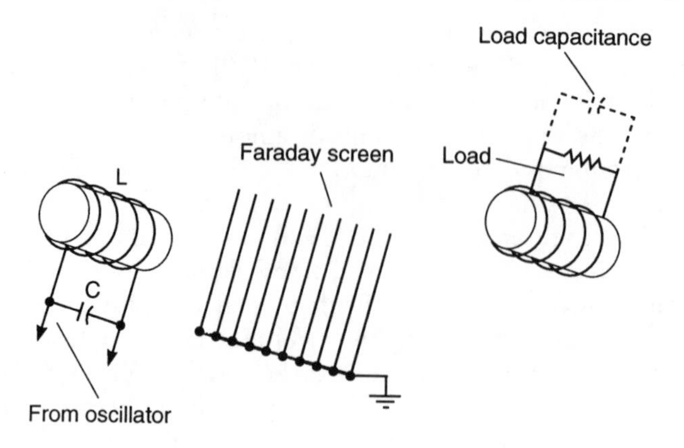

Fig. 3.22 *The Faraday shield*

higher frequency, capacitive coupling can be expected to transfer more harmonic power to the load than electromagnetic coupling.

Plate current is function of effective tank circuit Q

The plate current necessary to maintain oscillation in a Class-C feedback oscillator using an LC tank circuit as the frequency determining medium is approximately 1/Q times the alternating current in the inductive or capacitive branch of the tank. The no-load plate current of an oscillator with a high-Q tank is lower therefore than when a low-Q tank is used. By the same reasoning, we anticipate an increase in plate current when an oscillator is delivering power to a load, for loading decreases the operating Q of the tank. The loading characteristics of an LC feedback oscillator are shown in Fig. 3.23.

The need for load isolation

Let us suppose that we use a Faraday shield between oscillator tank and load coil and that the reactive component of the load is cancelled by tuning, that is by insertion of an equal reactance but of opposite kind. We would still find that the addition of the load affected the frequency of the oscillator. Why is this? We have, for simplicity, been using the resonance formula,

$$f_0 = \frac{1}{2\pi\sqrt{LC}}$$

For most of our discussion, this is justifiable when Q_0 is ten or greater. Inasmuch as oscillation frequencies often have to be maintained within

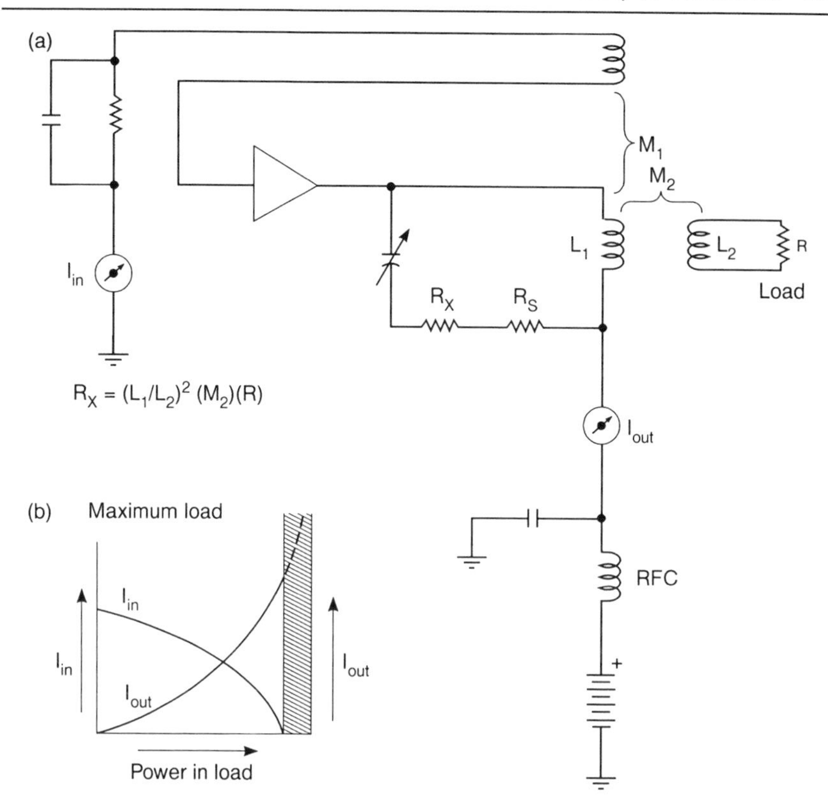

Fig. 3.23 *Loading characteristics of an LC feedback oscillator. a, Circuit: R_x is load reflected into LC tank. b, D.C. currents in oscillator. Oscillator no longer is operative in shaded region because of overloading and load power falls to zero*

relatively tight tolerances, it is well to appreciate that the mathematically rigorous relationship is:

$$f_0 = \left(\frac{1}{2\pi\ \sqrt{LC}}\right)\left(\sqrt{\frac{L - R_L^2\ C}{L - R_C^2\ C}}\right)$$

We see that the resonant frequency of a tank is affected by resistance as well as by the more influential inductance and capacitance. Thus, the resistance reflected into the tank by coupling the load to it changes the frequency of the oscillator. The resistance reflected into the tank also lowers its Q. The desired oscillation frequency can be restored by a tuning adjustment of the tank capacitor or inductor. The degradation in Q cannot be remedied for a given amount of power extracted from the oscillator. The decreased Q makes the oscillator frequency less stable with respect to changes in tube

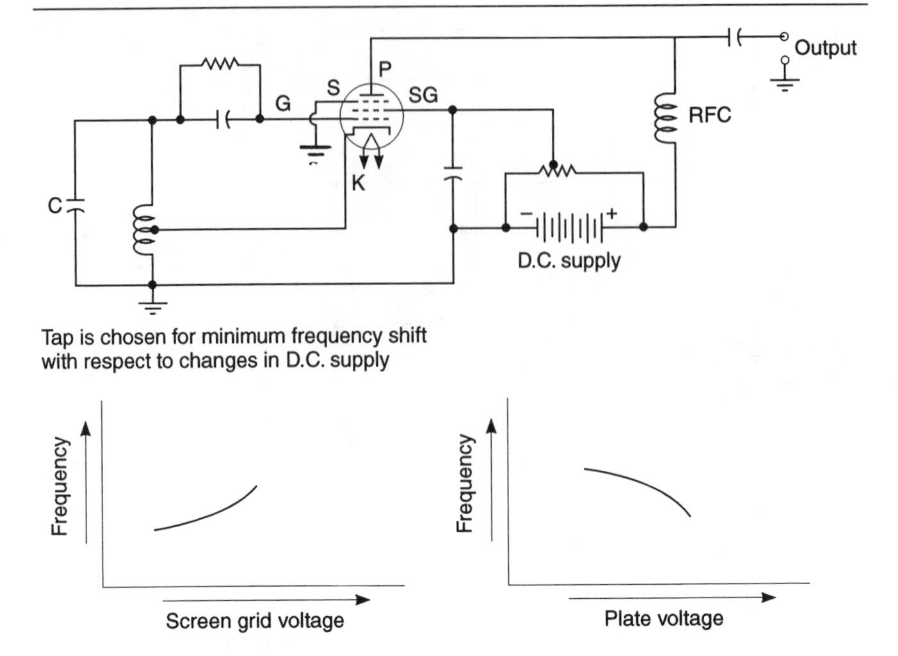

Fig. 3.24 *Basic electron-coupled oscillator and curves showing frequency variation*

characteristics, changes in operating voltages, the effects of temperature variations and mechanical disturbances. What is needed, we see, is a transfer of power from oscillator to load of such nature that the parameters of the load could not react upon the resonant tank. Such unilateral power transfer exists in the electron-coupled oscillator illustrated in Fig. 3.24.

The electron-coupled oscillator

The electron-coupled oscillator uses a tetrode or pentode tube in an arrangement in which the plate of the tube is *not* involved in the generation of oscillation. Rather, a complete oscillator circuit is formed with the cathode, control grid and screen grid. These elements act respectively as the cathode, grid and plate of an ordinary triode oscillator tube. We know that the screen grid in a tetrode or pentode tube intercepts a portion of the accelerated electrons from the cathode and that the remainder complete their journey to the plate. Those that reach the plate in the electron-coupled oscillator are velocity modulated at the oscillation frequency. This results in similar plate-current variations. Thus, a.c. power at the oscillation frequency is available in the plate circuit. When this power is dissipated in a load, a unilateral power transfer prevails; power is delivered from the screen grid (the 'plate' of the oscillator proper) to the tube plate, and the load circuit, but the load cannot appreciably influence the oscillator circuit. Neither the condition of maximum power of transfer, nor partial power transfer into a reactive load can exert much detuning, or Q-degrading, influence on the oscillator tank circuit. Some power amplification occurs between screen grid and plate. In this respect, the screen grid acts as a control grid of a triode amplifier. Indeed, the electron-coupled oscillator displays many of the features of an oscillator-amplifier arrangement. The screen grid is at a.c. ground potential. Therefore, it acts as an electrostatic shield, further isolating the oscillator and load circuits.

Circuitry considerations in the electron-coupled oscillator

The electron-coupled oscillator shown in Fig. 3.24 uses a Hartley oscillator in the cathode/control-grid/screen-grid circuit. The Hartley oscillator has not been discussed in this section. In an endeavour to explain concepts basic to all feedback oscillators, we have confined our examples to simple 'tickler' type feedback circuits. We see, however, that the Hartley configuration bears a similarity to the tickler feedback circuit. Essentially, a tapped coil rather than two separate coils is used. It is not the Hartley circuit in itself that is necessary to the electron-coupled oscillator. A tickler feedback circuit could be devised as well. The Hartley circuit as shown has the advantage that the screen grid is readily operated at a.c. ground potential. In the technical literature we also find electron-coupled oscillators with shunt-fed Hartley circuits in which the screen grid is 'hot'. This type of electron-coupled oscillator is satisfactory at low frequencies. Above several hundred kHz there may be excessive capacitive coupling between the screen grid and plate, thereby sacrificing the high degree of isolation attainable when the screen grid is operated at a.c. ground potential.

In order to avoid confusion, our electron-coupled oscillator is shown with a choke in its plate circuit. Sometimes a transformer, or, in very low power applications, a resistor, is used. Improved waveform purity results when a resonant tank tuned to the oscillation frequency is used. When this is done, it must be ascertained that the output is not due in part, or in entirety, to the 'tuned-plate tuned-grid' oscillator formed by having resonant tanks in the input and output circuits. Such malperformance is more probable at high than low frequencies, and is also more likely with tubes providing insufficient screen-grid shielding action.

Additional factors in the performance of the electron-coupled oscillator

When a pentode is used in the electron-coupled oscillator, it is generally best to ground the suppressor grid. This provides an additional Faraday shield between oscillator and load circuits. When the suppressor grid is grounded, there is less objection to an oscillator circuit in which the screen grid is above a.c. ground potential. When properly designed, the electron-coupled oscillator overcomes yet another trouble attendant to the ordinary self-excited oscillator. A change in d.c. plate voltage generally produces a shift in oscillation frequency in these circuits. In the electron-coupled oscillator, however, the screen-grid voltage and plate voltage changes resulting from a given change in d.c. power-supply voltage tend to shift the oscillation frequency in opposite directions. A wise choice in the ratio of plate to screen-grid voltage can result in high immunity to frequency change with respect to changes in the d.c. operating voltage from the power supply. The power output from the electron-coupled oscillator is limited for two reasons. First, the screen grid does not have sufficient power-handling capability to drive the 'amplifier' portion of the tube to high output. Secondly, the Class-C bias requirements of the *oscillator* section impose a limit on the output from the amplifier section. We see that the oscillator section functions as an elecrtron gun for the amplifier section, but the operation of both sections is very much determined by the automatic bias developed across the grid leak in the control-grid circuit.

4 Some practical aspects of various oscillators

In this chapter we will examine the characteristics of some of the basic feedback oscillator circuits. In so doing, we shall find ourselves in the advantageous position of being able to draw freely upon the fundamental concepts discussed in the preceding chapters.

Although oscillators such as the Hartley, the Colpitts, the tuned-output/tuned input and others can all be represented by a block diagram depicting the basic function of amplification, phase reversal, frequency determination and feedback apportioning, a close scrutiny of these oscillators reveals distinctly individual 'personalities'. It will now be our concern where the resonant tank is located, the actual nature of the feedback path, the means whereby the returned signal is caused to be in phase with the input signal, and other details which have been deemed of sufficient importance to merit the formal naming of specific oscillator circuitries. Hitherto, we have confined our discussions to 'tickler' feedback circuits in which inductive coupling between output and input coils in amplifiers established the feedback action. This type of oscillator sufficed as a model for analysing performance parameters common to all feedback oscillators. We will now proceed to other specific circuitries of the same family, i.e., LC feedback oscillators and closely related variants, such as crystal oscillators.

Three types of Hartley oscillators

The Hartley oscillator is one of the 'classical' LC feedback circuits. Unfortunately, the technical literature abounds with confusion with respect to the basic principles involved in the operation of this circuit. There are actually *three* mechanisms whereby oscillation may be produced in a Hartley oscillator. Although all three oscillatory modes are generally present, it will serve our objective best to consider each one separately. In this way, we will gain

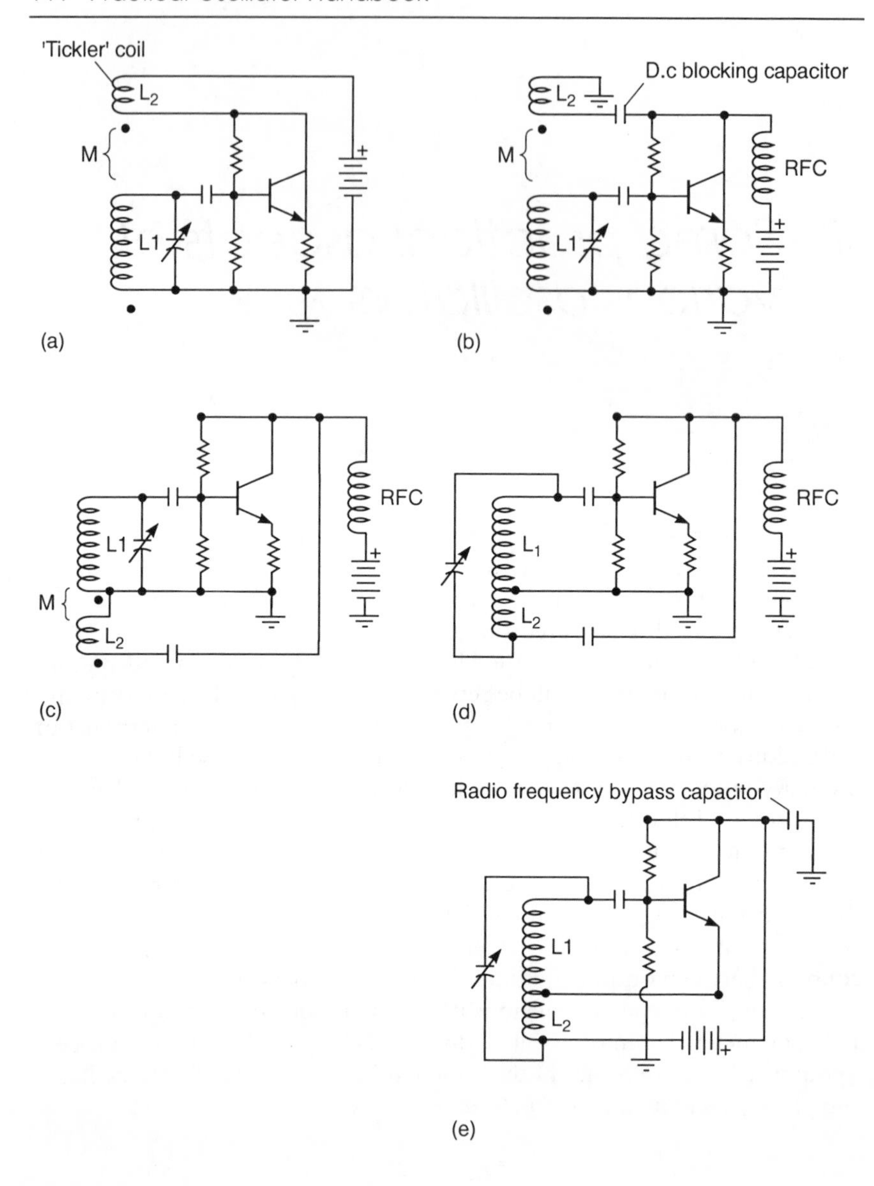

Fig. 4.1 *Evolution of Type-1 Hartley oscillators from basic tickler feedback circuit. a, Tuned base tickler oscillator. b, Tuned base tickler oscillator with shunt d.c. feed. c, Circuit b with physical rearrangement of tickler coil. d, Shunt-fed Hartley (hot collector) oscillator. e, Series-fed Hartley (hot emitter) oscillator*

practical insights into this sure-fire oscillator. The derivation or evolution of the Hartley oscillator from the basic 'tickler' feedback oscillator is depicted in Fig. 4.1. We will designate a Hartley oscillator with mutual-inductance tank-inductor sections, L_1 and L_2, as Type 1. A Hartley circuit without such coupling between L_1 and L_2 will be dealt with as Type 2. Finally, we shall consider as type 3, a configuration similar to that of Type 2, but with such small internal capacitance in the transistor that there is no chance of oscillation as the result of feedback through such internal capacity. It will be interesting to see whether such a 'stripped down' Hartley circuit can still oscillate.

The Type-1 Hartley oscillator

From Fig. 4.1 it is evident that the Type-1 Hartley is, indeed, very similar to the tickler-feedback circuit. What has been done is to substitute an auto-transformer arrangement in place of a conventional transformer in the feedback circuit. In both circuits feedback occurs because of inductive or electromagnetic coupling between the output and input sections of the oscillator. It is differences of this nature that endow the various oscillator circuits with unique personalities. To put the matter in more objective terms, some oscillator circuits are easier to implement in certain practical situations. In any event, the two Hartley circuits of Figs 4.1d and 4.1e are generally considered easier to build and place in operation than the tickler circuits. (A tapped-coil tends to be less monkey-business than two separate coils, and it is easier to get a high value for M, that is, tight coupling.) Summing up, the Type-1 Hartley oscillates because of electromagnetic coupling between output and input.

The Type-2 Hartley oscillator

Let us next consider the Type-2 Hartley circuit in which there is *no* mutual-inductance between the output and the input. As shown in Fig. 4.2, mutual-inductance, M, is zero. In practice, the coils could be far away from each other, or better, could be mounted so one is perpendicular to the other. Although the Hartley oscillator is not commonly constructed to attain this objective, it frequently happens that the coupling between L_1 and L_2 section of the tank is not as strong as it might be. How, then, can oscillation occur? One way in which the oscillator can still operate is via feedback provided by the internal collector-base capacitance of the transistor.

It may be argued that simply providing a signal path from collector to base results in negative, rather than positive feedback—it does not, in itself, produce the conditions required for oscillation. Recall, however, from Chapter 1, that an inductive load in the output circuit of an active-device can result in a negative input resistance. It was also pointed out that a

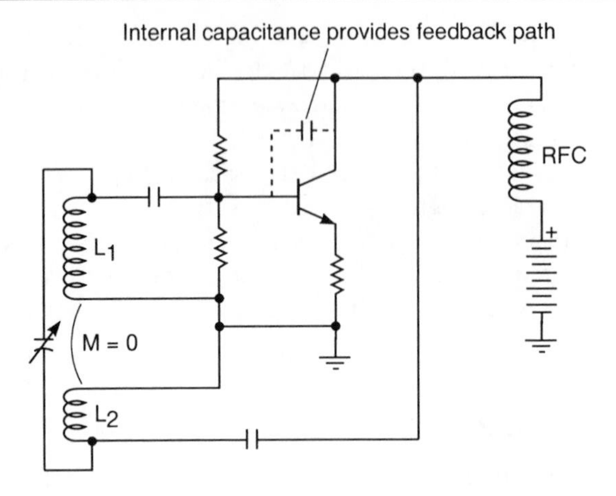

Fig. 4.2 *The Type-2 Hartley oscillator. Oscillation takes place despite the fact that there is no inductive coupling between the input and output circuits*

parallel-tuned tank circuit appears inductive when it is operated below its resonant frequency. We now see that our Type-2 Hartley circuit can oscillate even though there is no electromagnetic coupling between its output and input! This is so because at a frequency slightly below that corresponding to resonance of the tank circuit, the collector will 'see' an inductive load. The base circuit will then present a negative resistance to its tapped portion of the tank, thereby provoking it into oscillation.

The Type-3 Hartley oscillator

In the Type-3 Hartley oscillator, we postulate that, like the Type-2 circuit, there is no electromagnetic coupling between output and input; we further assume there is insufficient internal capacitance in the transistor to provide a feedback path. This implies that the negative-resistance mechanism described for the Type-2 circuit will not be present. (Of course, if the Type-3 Hartley *is* capable of oscillation, the tank circuit will still 'see' a negative resistance.)

The Type-3 Hartley circuit can be implemented with a tetrode or pentode tube, in which the internal capacitance is greatly reduced. However, we can simulate the operation under this condition by using the Type-2 transistor Hartley at a very-low frequency, say several hundred hertz. Such an arrangement is shown in Fig. 4.3. This circuit, can, indeed, be made to oscillate! The explanation is that the tank circuit now comprises a phase-shifting network—output energy is returned to the input, but displaced 180°. With sufficient amplifier-gain, such an arrangement will

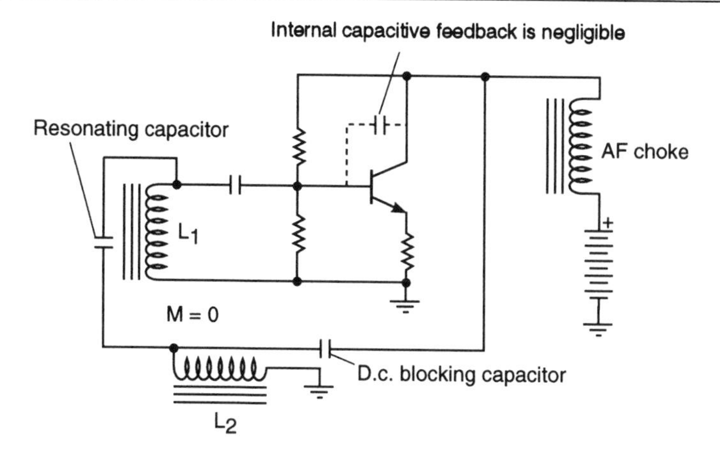

Fig. 4.3 *The Type-3 Hartley oscillator. Schematically the same as Fig. 4.2, this Hartley circuit, because of the low oscillation frequency, does not have sufficient collector-base capacitance to cause oscillation. Despite the lack of inductive coupling between output and input, this circuit* does *oscillate!*

oscillate, as witness the various phase-shift oscillators in which this condition is deliberately brought about via RC networks which are more lossy than LC circuits.

The oscillatory conditions generally found in practical Hartley circuits

Practical Hartley oscillators tend to operate with all three of the described oscillatory modes. That may well explain the reputation enjoyed by the Hartley oscillator as being a sure-fire circuit. It has been also found that Hartley oscillators tend to have very good loading characteristics, and have served well from very low to very-high frequencies. The tank inductor may be centre-tapped, but this much excitation is not usually needed. Somewhat more stable operation results from locating the tap closer to the ground or 'cold' end of the tank inductor. (This refers to the Hartley circuit of Fig. 4.1e.)

The Lampkin oscillator

The Lampkin oscillator is shown in Fig. 4.4. We see that the circuit is similar to the Hartley configuration, the essential difference being that the input connection is tapped down from the 'hot' capacitor-inductor junction of the tank. This minor change can, however, result in a considerable improvement in frequency stability. In Hartley, as well as other feedback

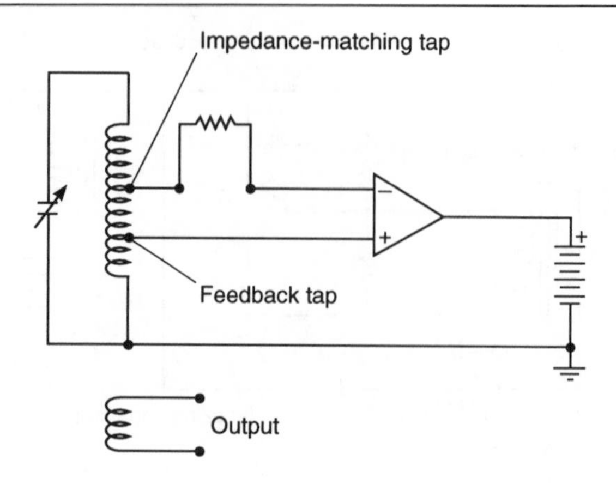

Fig. 4.4 *Basic Lampkin oscillator. One objective sought is to attain an approximate impedance match between the input of the amplifier and the resonant tank*

oscillators, which are often strongly self-excited, the input impedance of the active device can be lower than the impedance of the tank circuit connected to it. (A notable exception is the Clapp version of the Hartley.) As a consequence, the tank circuit impedance is shunted down and its Q is thereby lowered. In the Lampkin circuit, one technique is to drive the input of the active device from a portion of the tank that constitutes an approximate impedance match. Of course, the input impedance of the device may vary over the oscillation cycle, so the impedance matching concept cannot be considered in a strict mathematical sense. Nevertheless, we can say that the average value of the input impedance is more closely matched by this technique, than by simply driving the input from the 'top' of the tank circuit. Insofar as concerns frequency stability and preservation of the Q of the resonant tank, it is acceptable that the tap be located low enough down the tank so that the excitation occurs from a source having a *lower* impedance than that of the active-device input circuit. (The same objective, that of isolating the tank from the active device, is basic to the Meissner oscillator and is evident in other LC oscillators where the input resonant tank is transformed down in impedance.

In the Clapp oscillator, freedom from the shunting effect of the active device is achieved by using a series-resonant tank which inherently has a low impedance.) Although high mutual induction between the tapped portions of the tank inductor is not of primary importance in the ordinary Hartley, it is mandatory that close coupling exist between the inductor sections in Lampkin circuit. Otherwise, there will be a strong tendency toward oscillation at other than the desired frequency.

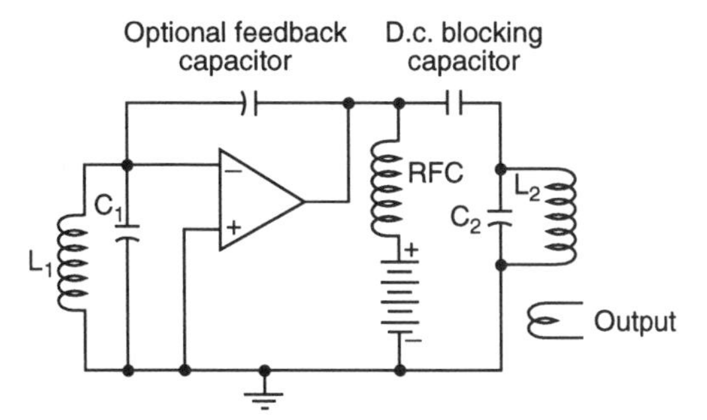

Fig. 4.5 *Tuned-output/tuned-input oscillator. In this shunt-fed circuit, the choke must not resonate with the LC tanks. (Because of their distributed capacitances, radio frequency chokes exhibit several series and parallel-type resonances)*

Tuned-output-tuned-input oscillator

Another classical LC feedback circuit stems from the Type-2 Hartley. If we tune the two tapped sections of the tank separately rather than employ one capacitor across the total tank turns, we arrive at the tuned-output/tuned-input oscillator (see Fig. 4.5). In the practical construction of this oscillator, care is exercised to *prevent* inductive coupling between output and input tank circuits. Oscillation then occurs as described in the Type-2 Hartley. The output tank operates at a frequency slightly below its resonant frequency, thereby presenting an inductive component to the output circuit. If a high-Q LC tank is connected to the input circuit, oscillation takes place at very nearly the resonant frequency of this tank. In this oscillator, the tank circuit with the highest Q tends to exert the preponderance of frequency control. Although this oscillator has traditionally been classified as a feedback oscillator, it would not be amiss to consider it as a negative-resistance oscillator. This, of course, is a subtle point. Any controversy provoked by such designation speaks well of the proponents of either argument. One of the central themes of this book is to stimulate thought concerning feedback and negative resistance. It is inevitable that insight into oscillator theory will result from speculations regarding the equivalences and differences between feedback and negative resistance. In any event, the tuned-output/tuned-input oscillator is not as persistent an oscillator as is the Hartley. However, due to the presence of two separate tuned circuits, this circuit tends to produce better wave purity than the Hartley. A well-screened tetrode or pentode will not work in this circuit without a small physical capacitor connected from plate to grid. When this is done, both plate and grid circuits must be made to appear inductive, in order to form a phase-inversion network. The action is not quite the same as that in a triode, but the difference is more academic than practical.

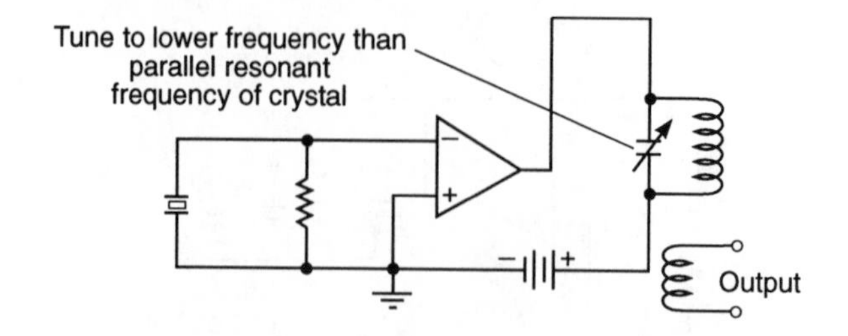

Fig. 4.6 *The Miller crystal oscillator. Although often referred to as a feedback oscillator, the crystal is provoked into oscillation because it 'sees' a* negative resistance *of appropriate magnitude*

The Miller oscillator

The Miller crystal oscillator (Fig. 4.6) is a tuned-output/tuned-input oscillator, in which a piezoelectric crystal is substituted for the input tank. The crystal operates very close to its parallel-resonant frequency. When a triode tube is used, oscillation occurs when the output circuit appears inductive inasmuch as the input circuit then presents a negative resistance to the crystal. When a well-screened tetrode or pentode is used, it is sometimes necessary to connect a 'gimmick', a very small capacitance, across the plate and grid terminals. The operation is then, for most practical purposes, identical with that obtained from a triode. The frequency will be very slightly lower, for in this way both tanks operate slightly below resonance. By so doing, the plate and grid inductances, with the capacitance connected from plate to grid, form a phase-inversion network. JFETs and MOSFETs behave like triode tubes.

At communications frequencies, most small pentode tubes will oscillate without the added external plate-grid capacitance. For given plate output levels, the pentode is to be preferred over the triode because the crystal is subjected to much less power. Both fracture and burning occur in crystals when the excitation is too high. In most applications where frequency control is the objective, crystal oscillators are operated at low power, on the order of a fraction of a watt to several watts, and drive an intermediate level (buffer) amplifier, or excite the power output amplifier directly. When the Miller crystal oscillator is designed around a junction transistor, it is often necessary to select a high activity crystal, due to the drastic impedance mismatch between the parallel-resonant crystal and the input of the transistor. In other respects, operation can be compared to its vacuum-tube counterpart.

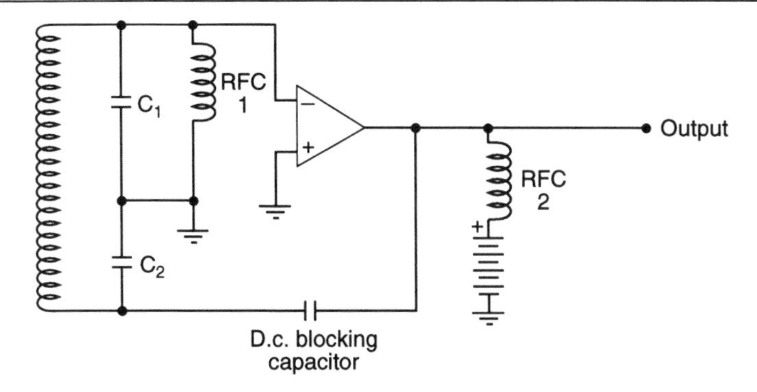

Fig. 4.7 *The shunt-fed Colpitts oscillator. The capacitive divider formed by C_1 and C_2 is the hallmark of the Colpitts circuit. The quality of the radio frequency chokes is important in the shunt-fed version. RFC_1 is usually optional; if used, the two chokes should be different types to avoid an undesired resonance*

The Colpitts oscillator

The Colpitts oscillator is the third of the 'classical' LC feedback circuits. The two common versions of the Colpitts oscillator are shown in Figs 4.7 and 4.8. The configuration of this oscillator resembles that of the Hartley. The difference is that the tapped connection is made at *capacitor* junctions rather than at *inductor* junctions. As a result, the tank circuit forms a phase-inverting network. The Colpitts cannot operate in an analogous way to the Type-1 Hartley because the electrostatic equivalent of electromagnetic induction is not present. Operation is most like that occurring in the Type-3 Hartley wherein the phase-inverting properties of the LC network make the most direct contribution to oscillation. The Colpitts circuit generates a frequency slightly *higher* than the resonance of the tank circuit. It is indeed under such a condition that phase inversion occurs between the output and input circuits of the active device.

The two capacitors constitute a *voltage divider* and their ratio governs the excitation of the oscillator. Sometimes we find these capacitors ganged, in which case constant excitation can be maintained over a fairly wide frequency range. Another tuning method that can attain similar ends is to employ fixed capacitors for C_1 and C_2 and a variable capacitor across the entire coil. This method is most satisfactory when the desired frequency tuning-range is moderate. For relatively-small tuning excursions, a vernier capacitor across either C_1 or C_2 is satisfactory.

The salient feature of the Colpitts is its comparatively good wave purity. This is due to the fact that C_1 and C_2 provide low-impedance paths for harmonics, effectively shorting them insofar as concerns the input of the

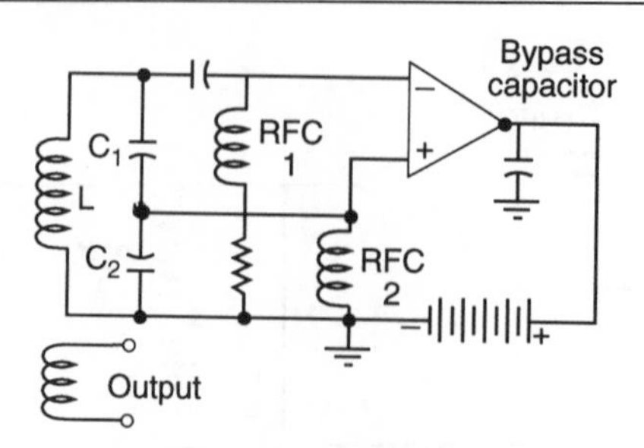

Fig. 4.8 *The Colpitts oscillator (series feed). Here again, RFC_1 is usually optional. If used, the two chokes should be different types*

active device. Another noteworthy feature of the Colpitts is that it is an exceptionally fine performer at high frequencies, well into the microwave region. (The active device must, of course, have high-frequency capability.) Initially C_1 and C_2 can be equal. Excitation can then be reduced by making C_2 larger than C_1.

Another type of Colpitts oscillator, with a crystal substituted for the inductor of the parallel-tuned tank, is depicted in Fig. 4.9. Because of the crystal-holder capacitance, some experimentation with C_1 and C_2 may be required in order to obtain sufficient self-excitation. In this circuit, the crystal operates at slightly below its parallel-resonant frequency and thereby simulates an inductance.

The ultra-audion oscillator

The ultra-audion circuit can be shown to derive from the basic Colpitts oscillator, characterized by the capacitive voltage divider in the feedback path. However, the reliance is on internal capacitances of the active device, rather than external physical capacitors. This can be clearly seen from inspection of the circuits of Fig. 4.10. In practice, some of these internal capacitances can be attributed to the package of the device. In any event, they are relatively small and are at least a partial explanation of the good high-frequency performance of this oscillator circuit. With various minor modifications, it enjoys popularity at VHF, UHF and in the microwave portion of the spectrum. As its name implies, its lineage traces to electron-tube technology. It tends to be a sure-fire oscillator, providing the active device has adequate high-frequency capability. Otherwise, if operation is not forthcoming, the radio frequency choke should be investigated; at high frequencies, chokes often display unfortunate series-type resonances as well as the expected parallel resonances. Often, a non-inductive resistor may be a better choice. JFETs and MOS devices work very well in this circuit.

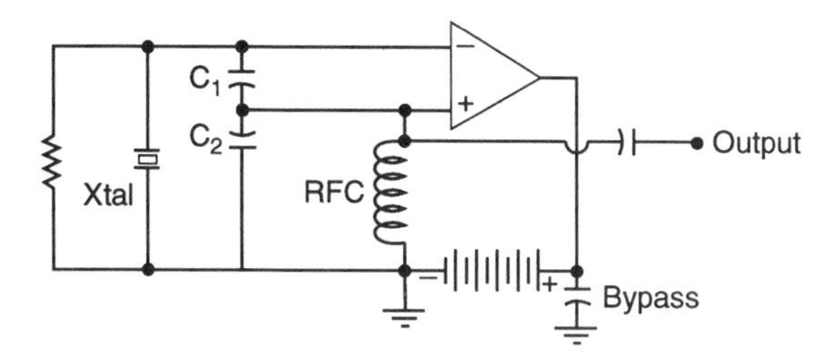

Fig. 4.9 *Colpitts oscillator with crystal substituted for parallel-tuned tank*

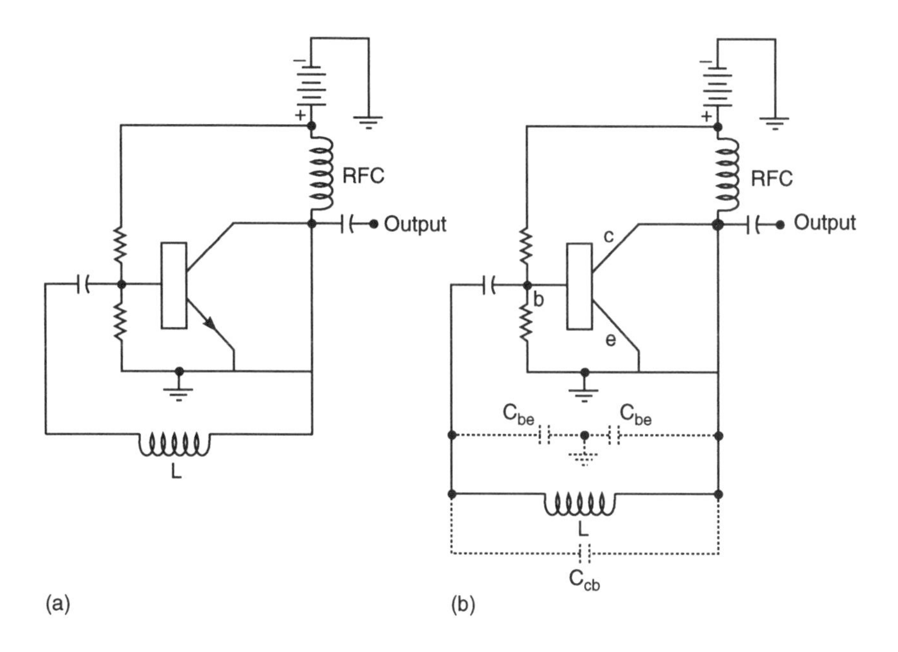

Fig. 4.10 *The ultra-audion oscillator: a member of the Colpitts family. a, Basic hardware circuit. b, The equivalent circuit. The configuration of the internal capacitances form a Colpitts-type oscillator*

The Pierce oscillator

The Pierce oscillator is a novel circuit in that no LC resonant tanks are employed. We see in Fig. 4.11 that a crystal is connected from output to input of the active device. At a frequency slightly higher than that corresponding to the series-resonant mode of the crystal, the crystal will appear inductive. This simulated inductance forms, in conjunction with the interelectrode capacitances, a Colpitts tank network (see Fig. 4.12). It might be argued that a similar circuitry condition would exist at a frequency

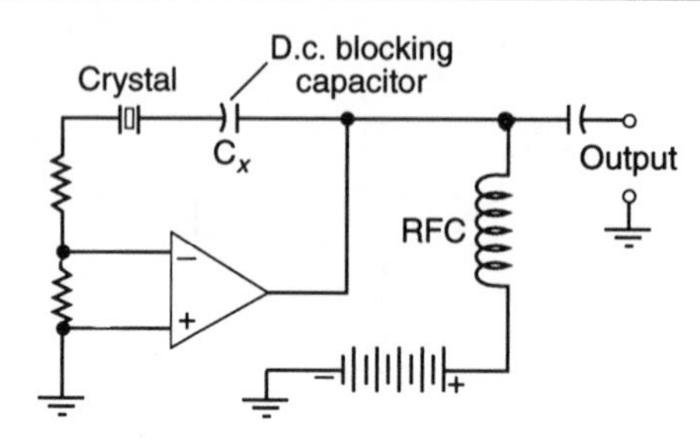

Fig. 4.11 *The Pierce crystal oscillator*

slightly lower than the parallel-resonant frequency of the crystal. This is true, but the equivalent impedance in the vicinity of the series-resonant mode is much less than in the neighbourhood of parallel resonance. As a consequence, feedback is greater near series resonance and the circuit finds this mode more favourable for oscillation. Often it is found necessary to connect a small capacitor between the input terminal and ground to bring about oscillation. When such a capacitor is variable, a vernier tuning range is obtained. Sometimes a parallel-tuned tank is inserted in the output in place of the choke in order to improve wave purity. When this is done, the tank is operated at a frequency slightly higher than that corresponding to its resonance, in order to present a capacitive component to the output circuit. Such an operating condition is automatically attained by simply tuning for maximum output. Fragile crystals can be easily damaged in this oscillator if the power level is too high.

The Clapp oscillator

The Clapp oscillator is another Colpitts derivative. The distinguishing feature of this circuit is that a series-tuned tank is employed as part of the total feedback network (see Fig. 4.13). Oscillation occurs at a frequency slightly higher than the resonant frequency of the series-tuned tank. Under this condition, the series-tuned tank appears inductive. Thus, an LC phase inversion network is formed in conjunction with voltage-divider capacitors C_1 and C_2 which is of the same basic configuration as the conventional Colpitts tank circuit. Due to the fact that the series-tuned portion of the tank has a much lower impedance than parallel-tuned tanks, it is postulated that the Clapp oscillator tank circuit is relatively immune to variations in tube parameters and loading conditions. This, it is contended by those favouring this oscillator, naturally manifests itself as excellent frequency stability.

There is much truth in the above assertion. However, other oscillator circuits also exhibit good performance when care is taken to limit excitation, when high-Q tanks are employed, and when temperature-compensating

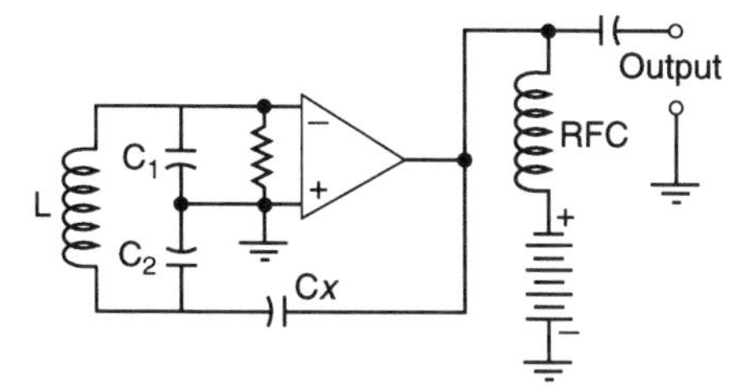

Fig. 4.12 *Equivalent circuit of the Pierce crystal oscillator. C_1 and C_2 are internal capacitances of the active device. For a transistor, C_1 represents emitter-to-base capacitance; C_2 is the emitter-to-collector capacitance. For other devices, similar correspondences prevail. Note the Colpitts format*

capacitors are inserted in the resonant circuits. The Clapp oscillator invariably embodies these precautions, and often incorporates electron-coupled output. This is because this oscillator has been employed primarily for governing the frequency of amateur transmitters. Its history has evolved very little in the way of workhorse applications. Conversely, the other oscillator circuits, primarily because of their readiness to oscillate, have often been designed and constructed in ways not conducive to optimum frequency stability. Although there is no magic shortcut to frequency stability in the Clapp oscillator, it does appear to be relatively easy to construct and place into satisfactory operation. Generally, C_3 will be much smaller than either C_1 or C_2. (For high Q, the tank needs a high L to C ratio.)

The tri-tet oscillator

The tri-tet oscillator (Fig. 4.14) is a circuit particularly well suited for the generation of selected harmonics of a crystal's fundamental oscillatory frequency. Such frequency multiplication is often desirable because quartz crystals tend to become mechanically fragile, electrically inactive and expensive at frequencies above, say, 10 MHz. An additional advantage of harmonic operation is that harmonically related communications frequencies may be readily selected; these harmonics have the same percentage frequency stability as the fundamental frequency. The name 'tri-tet' derives from the simultaneous operation of the circuit as a triode oscillator and a tetrode amplifier. In this respect the tri-tet circuit resembles the electron-coupled oscillator. However, isolation between oscillator and amplifier portions of the circuit is dependent upon the fact that the amplifier is tuned to a harmonic of the oscillation frequency.

The isolation between load and oscillator, which would be otherwise

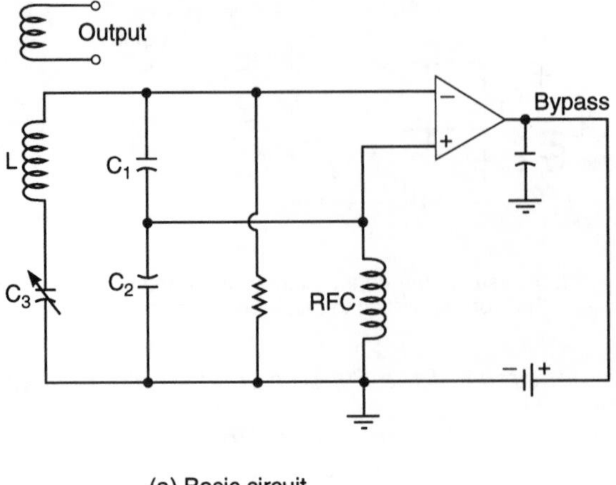

(a) Basic circuit

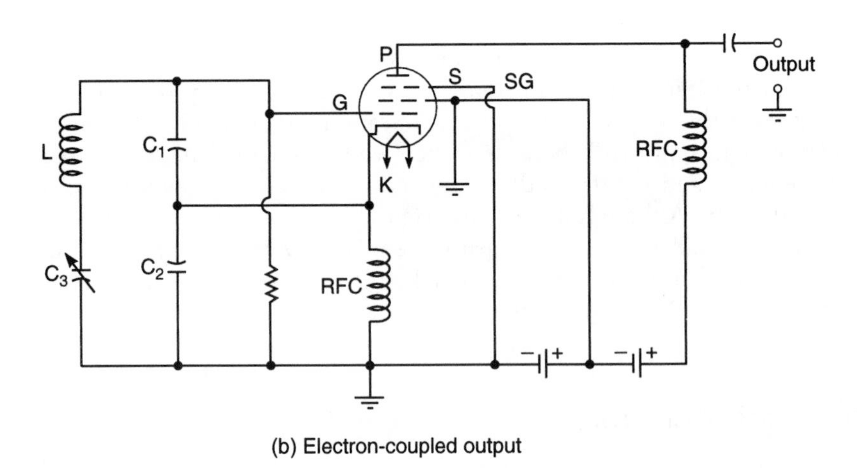

(b) Electron-coupled output

Fig. 4.13 *Clapp oscillators*

obtained from electron coupling, is negated by the common impedance, L_1C_1, inserted in the cathode lead. The oscillator section of the tri-tet circuit is a Type-3 Hartley in which the crystal acts as one portion of the total tank inductance and the parallel-tuned tank, L_1C_1, constitutes the other. In order for this to be so, the oscillation frequency is slightly below the parallel-resonant frequency of the crystal. For the same reason the oscillation frequency must be below the parallel-resonant frequency of tank L_1C_1. Indeed, increased harmonic output results when this tank is actually resonated to the desired harmonic. Output tank L_2C_2 is always resonated at the desired harmonic frequency.

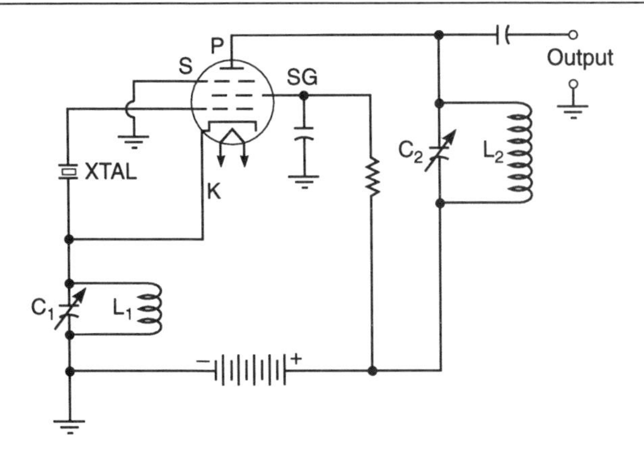

Fig. 4.14 *The tri-tet oscillator*

When the output tank, L_2C_2, is tuned for fundamental frequency operation, the crystal current is likely to become dangerously high. This danger can be eliminated in two ways. First, tank L_1C_1 can be tuned to a very much high frequency than the fundamental. In actual practice, L_1C_1 could remain tuned to one of the harmonics. Secondly, tank L_1C_1 can be short-circuited. The circuit will then operate as a Miller or tuned-plate/tuned-grid oscillator because at radio frequencies there is usually sufficient internal feedback within the tube. If a well-screened tube is used at low and medium radio frequencies, it may be necessary to connect a gimmick capacitance between plate and grid terminals for fundamental frequency oscillation when tank L_1C_1 is short-circuited. The harmonic generating ability of the tri-tet oscillator is directly related to its excitation; that is, it must operate deeply in the Class-C region in order that grid and plate current pulses are steep and of short duration (rich in harmonic content). This condition is enhanced by using an active crystal, and a high quality radio frequency choke in the grid circuit. It should be appreciated, however, that a so-called 'harmonic' crystal is not necessary in this circuit.

The Meissner oscillator

The Meissner oscillator employs electromagnetic feedback, but there is no conductive connection between the resonant tank and the remainder of the circuit. In Fig. 4.15 there is no intentional mutual induction between L_1 and L_2. (In practical Meissner oscillators, the coupling between these coils is, at least, so loose that no oscillation is thereby produced.) There is, however, mutual induction between the inductor of the resonant tank and these coils. Oscillation occurs at approximately the resonant frequency of the tank, because feedback then attains its maximum value. The coupling between

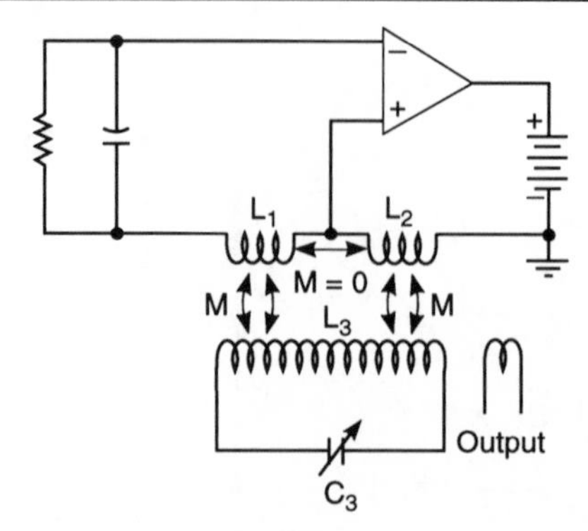

Fig. 4.15 *The Meissner oscillator. The basic idea is that feedback from L_2 to L_1 must not occur directly, but rather couples to L_3*

the three circuits should be no greater than that which is necessary for oscillation with a moderate load. (The load can be inductively coupled to L_3.) Otherwise, this circuit is likely to exhibit a pronounced tendency to generate spurious frequencies. This is so because stray and distributed capacitances associated with L_1 and L_2 resonate these coils at frequencies other than those corresponding to tank-circuit resonances; there is always transformer-type coupling between the three coils regardless of the tuning of C_3. It is highly desirable to use a Faraday shield between the tank coil, L_3, and the feedback coils, L_1 and L_2. This greatly reduces the possibility of oscillation at spurious frequencies due to capacitive feedback. This oscillator is uniquely suited for certain instrumentation and measurement techniques, due to the fact that the resonant tank merely has to be brought in proximity with the electronic circuit proper. Under laboratory conditions, the Meissner oscillator may be adjusted to operate so that there is a relatively high isolation between the active device and the resonant tank. For practical applications, the Meissner oscillator is not as popular as the other LC feedback circuits.

The Meacham bridge oscillator

The Meacham bridge oscillator consists of a high-gain 'flat' amplifier, in conjunction with a Wheatstone bridge containing a resonant arm (see Fig. 4.16). The bridge is inserted in a positive feedback loop between the output and input of the amplifier. The resonant arm of the bridge is generally a crystal. When the crystal is operated at its series-resonant mode, its effective resistance then tends to balance the bridge, thereby producing extreme (theoretically infinite) attenuation of the feedback signal. Under such a condition, oscillation could not exist. It is clear that a slight imbalance of the

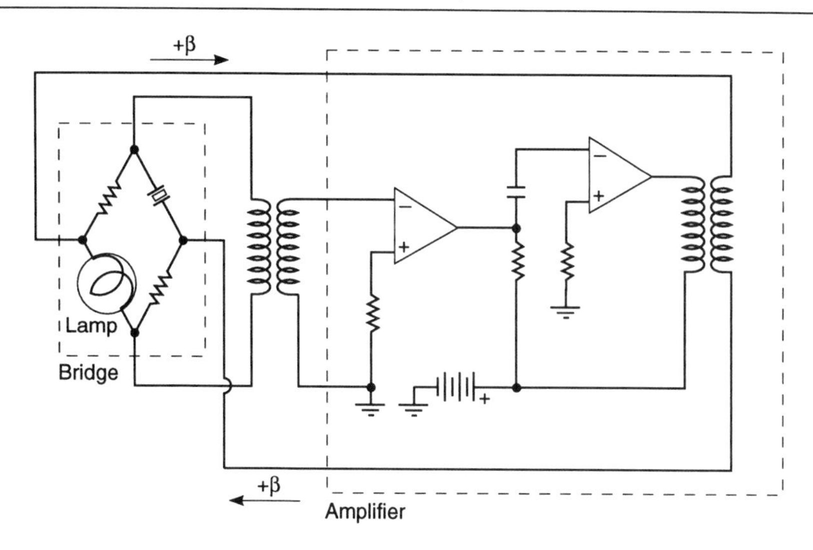

Fig. 4.16 *The Meacham bridge oscillator. Operating very close to balance, the bridge automatically regulates feedback, thereby maintaining Class-A amplifier operation, exceptional frequency stability and good wave purity*

bridge is necessary to enable the circuit to meet the criterion of oscillation. It is not desired that this imbalance be attained by off-resonant operation of the crystal, for the phase shift thereby produced would again defeat the requirements of the oscillations criterion. If the phase were corrected elsewhere in the circuit, such operation would not advantageously make use of the optimum frequency stability that can be provided by the crystal. Bridge imbalance is attained by using a small tungsten filament lamp as one of the bridge arms. Such a lamp has a positive temperature coefficient; its resistance increases with operating temperature.

Let us suppose that the lamp is cold and the circuit has just been turned on. The bridge is grossly out of balance, enabling a strong feedback signal to be impressed at the input of the first amplifier stage. The resulting oscillatory current heats the lamp filament; the attendant increase in lamp resistance causes the bridge to approach its balanced condition. Feedback is accordingly reduced until equilibrium is attained with the bridge imbalanced just sufficiently to permit oscillation. Any tendency for the oscillatory amplitude to change is counteracted by a change in lamp resistance, which shifts bridge balance in the direction required to oppose the change. We see that the bridge not only governs oscillation frequency, but also provides amplitude stabilization. The amplifier operates in its Class-A region, for in the Meacham bridge oscillator the performance objective is frequency stability, not power. Other things being equal, frequency stability is generally enhanced by good wave purity and by amplitude stabilization. An outstanding feature of the Meacham bridge oscillator is that the bridge circuit operates with an effective Q several times higher than that of the crystal itself. This is due to the fact that a near-balanced Wheatstone bridge displays a phase-magnifying property.

In Fig. 4.16 we see '$+\beta$' symbols. It should not be inferred from this that there are two positive feedback paths. A feedback path in any feedback oscillator involves a complete circuit or loop. Most often, one portion of such a loop is the grounded part of the circuit. This makes it schematically convenient simply to designate the 'hot' connection, or connections, with the $+\beta$ symbol, it being then understood that the ground side of the feedback loop also participates in the feedback of energy from the output of the oscillator to the input.

In the basic Meacham oscillator of Fig. 4.16, neither side of the external feedback path is grounded. The two $+\beta$ symbols merely indicate the route whereby occurs one complete go and return path, or loop for the return of energy from the oscillator output to its input circuit.

In the event one of the two connections designated by $+\beta$ signs in Fig. 4.16 were grounded, the basic principle of oscillation would not be altered. In such a case, it would be conventional to label only the ungrounded feedback connection with the $+\beta$ sign. Whether or not such grounding would be employed in a practical Meacham oscillator would depend very much upon the kind of electrostatic shielding used in the bridge components and the associated transformers. The basic idea here would be to avoid imbalancing the bridge in such a way that the frequency-governing effect of the crystal would be impaired. In most instances, this matter would be of little consequence for frequencies up to the middle audio range of the spectrum.

Line oscillators

Line oscillators are advantageously used at frequencies sufficiently high to enable practical exploitation of the resonant properties of transmission lines. A guidepost here is that the wavelength corresponding to a frequency of 300 MHz is 1 m. A quarter-wave line for this frequency would therefore be approximately 9 inches in length. Resonant lines associated with practical UHF oscillators are always physically shorter than the theoretical free-space wavelength. In the first place, the velocity of electromagnetic wave propagation along the line is slightly less than in free space. This is a relatively small effect in designing line oscillators, however. The major shortening effect is produced by the action of tube and stray capacitances across the high-impedance end of the resonant line. In Fig. 4.17 we see a simple line oscillator, which resembles the low-frequency version (using lumped tank

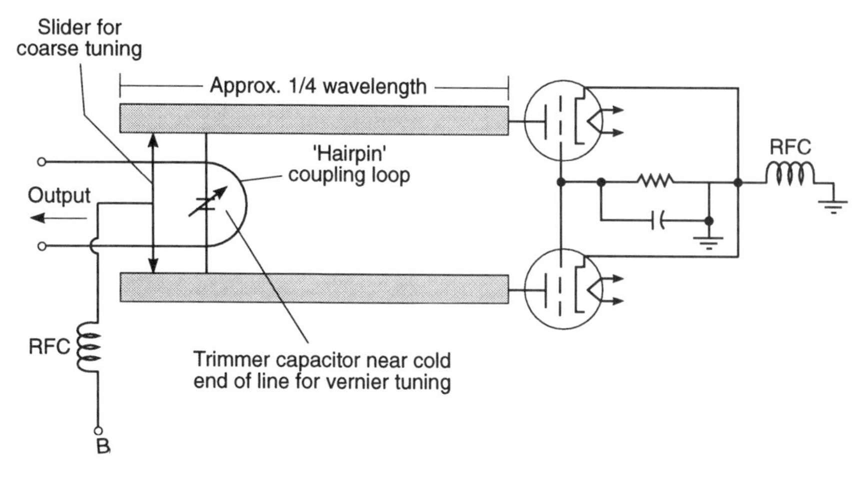

Fig. 4.17 *A line oscillator for UHF*

circuit) of a push–pull ultra-audion oscillator. The exact oscillator type is often obscure in such UHF oscillators because there is sufficient 'radiation coupling' between the resonant lines and the tube elements. UHF output energy is extracted from the cold end of the resonant line by means of a half-turn 'hairpin' loop electromagnetically coupled to the line. The impedance of such a loop is low enough to feed directly into coaxial cable. Vernier tuning of this oscillator is provided by a small trimmer capacitor connected near the shorted end of the line.

Another line oscillator

Another line oscillator is shown in Fig. 4.18. Basically a tuned-plate/tuned-grid circuit, frequency control is provided by a quarter-wave section of coaxial line. The inductive reactance needed in the plate circuit for oscillation results from operation of the conventional 'lumped' LC tank slightly below parallel resonance. Somewhat improved operation may result from the use of a butterfly resonator in the plate circuit (a tuning unit in which inductance and capacitance are integrally incorporated in a single mechanical assembly). However, frequency control is primarily exerted by the quarter-wave coaxial line in the grid circuit. Such a line may be constructed to have an extremely high Q. Theoretically Q is maximum when the ratio of outer conductor diameter to inner conductor diameter is approximately 3.6 to 1. Practically, ratios between 3 and 5 appear to be quite satisfactory. Temperature effects may be overcome by constructing the line of alloys such as Invar or Monel metal which have a low coefficient of expansion. In order to prevent degradation of the high Q inherent in such a structure, the inner conductor should be tapped as close to the shorted end of the line as is

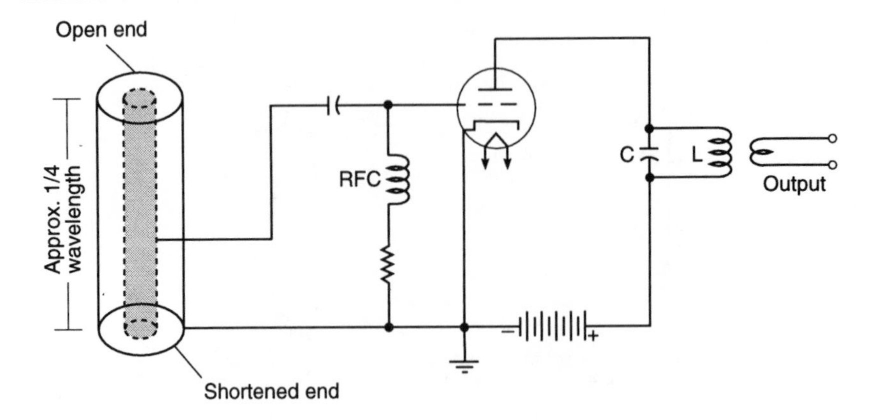

Fig. 4.18 *Line oscillator employing high-Q quarter-wave coaxial line for frequency control*

consistent with oscillation. The coaxial line is inherently shielded and does not radiate as does the open type two-conductor line. This avoids a dissipative loss and also makes the resonant frequency immune to proximity effects of nearby objects.

The magnetostriction oscillator

The magnetostriction oscillator makes use of electromechanical resonance of magnetostrictive rods as described in Chapter 1. It will be seen from Fig. 4.19 that the circuit configuration is suggestive of the Hartley oscillator. However, the phasing of the two windings is opposite to that required to produce oscillation in the Hartley circuit. As a result, there is sufficient negative feedback to discourage Hartley type oscillation. Oscillation occurs at the *mechanical resonant frequency* of the magnetostrictive rod, for only then is voltage phased for positive feedback induced in the grid-cathode winding. An important aspect of the circuit of Fig. 4.19 is that d.c. plate current flows in one winding. This establishes an operating bias about which oscillation occurs. The oscillation frequency is:

$$\frac{\text{speed of sound in rod}}{2 \times \text{length of rod}}$$

The same units of length should be used in both numerator and denominator of this relationship. If the speed of sound is designated as so many unit lengths per second, the frequency is obtained as so many hertz. For example, the speed of sound in a certain nickel alloy may be of the order of 4800 meters per second. A rod one meter in length would then oscillate at the

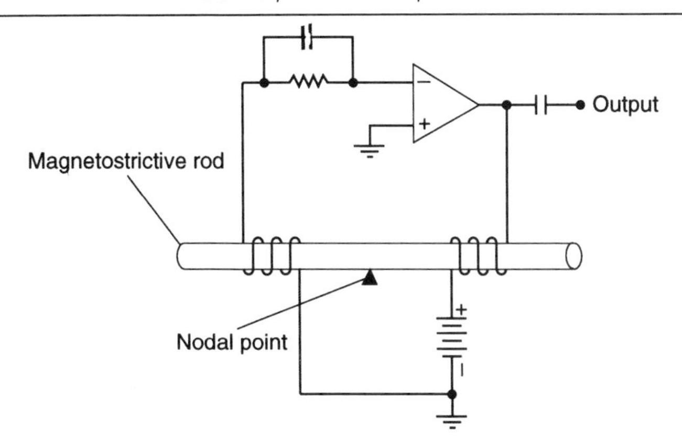

Fig. 4.19 *The magnetostrictive oscillator. Despite the topographical resemblance to a Hartley circuit, the oscillation frequency is governed by the magnetostrictive rod not by LC relationships. Also, the phasing of the windings is such as to prevent Hartley oscillation*

frequency of 2400 Hz. If the circuit were arranged so that no magnetic bias was provided, the rod would tend to vibrate at twice the frequency of the exciting current because contraction results from electromagnetic forces of both polarities. Such operation is not very efficient. In order to avoid damping the effective Q of the vibrating rod, it must be clamped at its nodal point. This is the point or narrow region where vibrational amplitude is minimum. Another technique relevant to increasing the Q is to use a laminated bundle of thin strips of material rather than a solid rod.

The Franklin oscillator

The Franklin oscillator uses a two-stage amplifier in conjunction with a parallel-tuned resonant tank circuit (see Fig. 4.20). In this circuit, each stage shifts phase 180° so that the total phase shift is 360°, which is equivalent to zero phase shift. We may say that one stage serves as the phase inverting element in place of the RC or LC network which generally performs this function. It is, from an analytical viewpoint, immaterial which stage we choose to look upon as amplifier or phase inverter. The configuration is essentially symmetrical in this respect; *both* stages provide amplification and phase inversion. The salient feature of the Franklin oscillator is that the tremendous amplification enables operation with very small coupling to the resonant circuit. Therefore, the frequency is relatively little influenced by changes in the active device, and the Q of the resonant circuit is substantially free from degradation. The closest approach to the high-frequency stability inherent in this oscillator is attained by restriction of operation to, or near to,

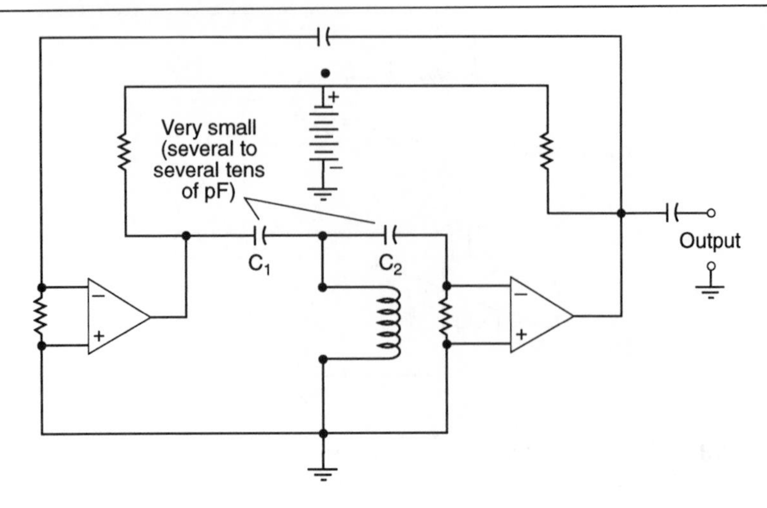

Fig. 4.20 *Basic Franklin oscillator*

the Class-A region. This should not be accomplished by lowering the amplification of the two stages, but, rather by making the capacitors C_1 and C_2 very small. Additionally, a voltage-follower 'buffer' stage is helpful in this regard. Extraction of energy directly from the resonant tank, would, of course, definitely negate the frequency stability otherwise attainable. Obviously, the Franklin oscillator is intended as a low-power frequency-governing stage, not as a power oscillator. The feedback capacitor at the top of the circuit of Fig. 4.20 is not critical; it can be about ten times the capacitance of C_1 or C_2.

The Butler oscillator

The Butler oscillator uses a two-stage amplifier with a crystal inserted in the positive-feedback loop embracing the entire amplifier (see Fig. 4.21). In tube parlance, we have a 'grounded grid' stage driving a cathode follower. Feedback is derived from the cathode of the output stage and returned to the cathode of the input stage. This feedback connection provides the phase conditions for oscillation without need for phase-shift networks. Accordingly, the crystal oscillates at the frequency at which feedback is maximum. This occurs at its series-resonant frequency because its impedance is then minimum. Indeed, the principal use of the Butler oscillator is found in timing equipment where attaining oscillation at a precise frequency is desired. When the crystal is employed in other oscillator circuits, it generally operates at neither the exact series-nor parallel-resonant frequency. The Butler oscillator is a very convenient circuit to construct with a tube

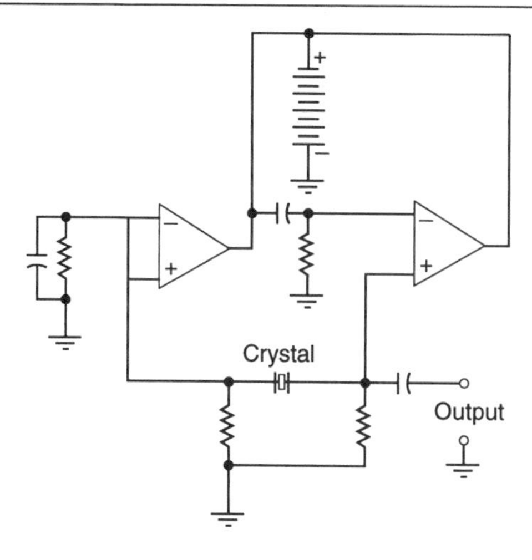

Fig. 4.21 *Basic Butler oscillator*

inasmuch as a duo-triode or tiode-pentode within one envelope is admirably suited to the configuration. When it is desired to operate the crystal off-resonance, a vernier tuning range may be obtained by connecting a variable capacitor in series with the crystal. Output may also be derived from the output circuit of stage 2 by inserting a resistor or choke in the d.c. supply. In any event, optimum results will be obtained when a buffer or isolating amplifier is employed. Transistor arrays are available with several independent devices in one package.

Bipolar transistor oscillators

Practically any electron-tube oscillator can be simulated by a transistorized version of the basic configuration. (A possible exception is the electron-coupled oscillator.) Once the d.c. biasing arrangement has been given appropriate attention, the transistor is a near substitute for the three element tube. For Class-A amplifiers the relatively low input impedance of the transistor contrasts violently with the extremely high input impedance of the electron tube. However, in Class-B and C oscillators, both transistor and electron tube circuits can display a low input impedance. Actually, the transistor is more suggestive of the pentode tube, notwithstanding the discrepancy in the number of elements. This is so because the collector circuit behaves as a constant-current generator, as does the plate circuit of the pentode. Both output circuits tend to have relatively high impedance.

We see in Fig. 4.22 typical transistor versions of popular LC oscillators. In the ultra-audion type Colpitts, the internal emitter to base capacitance

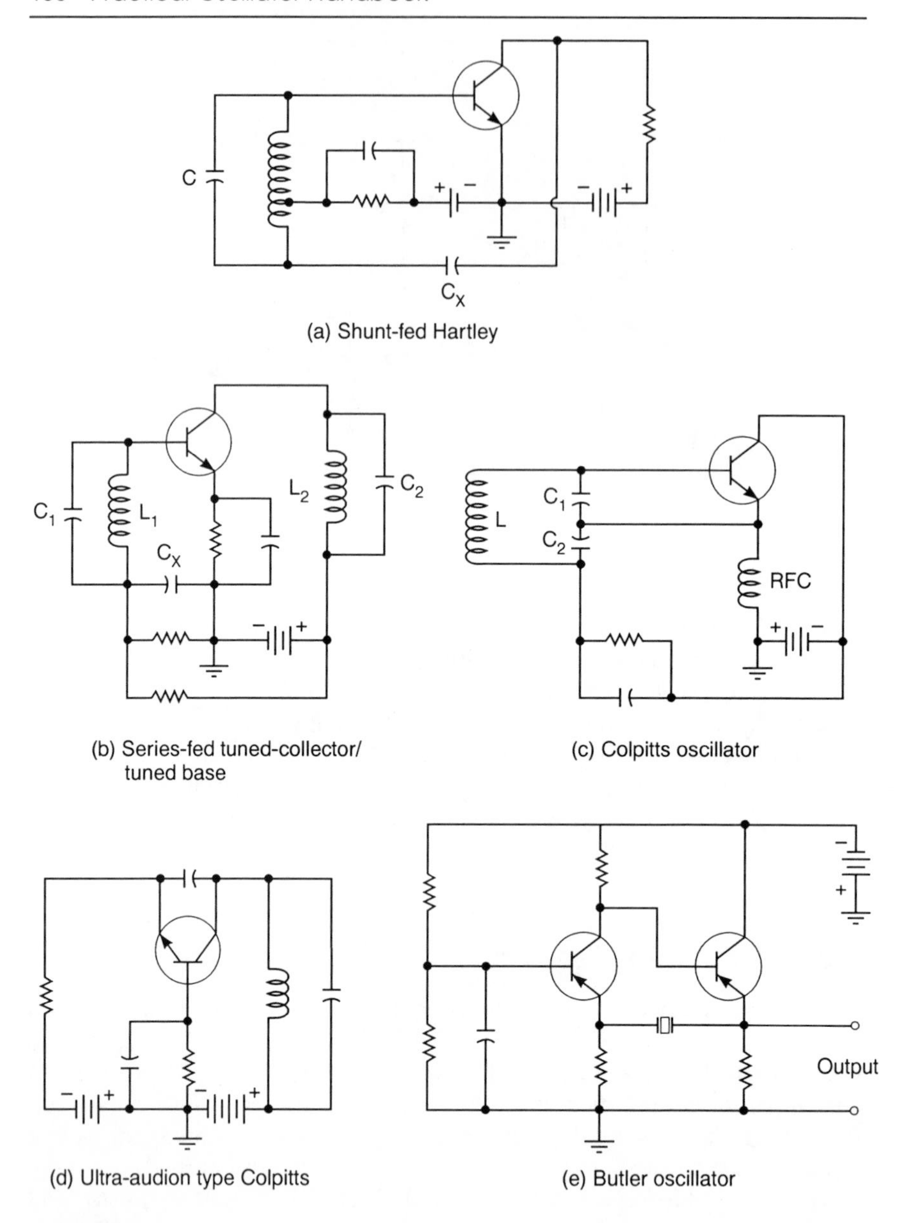

Fig. 4.22 *Representative oscillators using bipolar transistors. Usually a bit of experimentation will pay dividends in optimizing operation for stability, self-starting, wave purity, power output, etc. D is another version of the circuit of Fig. 4.10*

constitutes the second capacitor of the Colpitts capacitive voltage divider. In the tuned-output/tuned-input oscillator, the transistor exhibits a 'Miller' effect; that is, its input acts as a negative resistance when the output tank is tuned to appear inductive at the parallel-resonant frequency of the emitter-base tank. Although transit-time effects and interelectode capacitance ultimately impose high-frequency performance limits in electron tubes, a wide frequency spectrum is permissible before attention must be given to these restricting factors. With transistors, however, even at moderate radio frequencies, it is essential to ascertain that the current amplification of the particular transistor type is compatible with the desired oscillation frequency. Germanium power transistors designed for several tens of watts power dissipation rapidly lose current amplification beyond several kilohertz. In contrast, silicon transistors of the diffused base or 'mesa' type are suitable for oscillators generating from several tens to several hundreds of megahertz. Oscillation in the gigahertz range can be obtained from appropriately-packed silicon and gallium-arsenide transistors made for microwave applications.

The unijunction transistor oscillator

The unijunction transistor is a switching device in which conduction is abruptly established across two of its three terminals when the voltage impressed across those terminals attains a certain value. Once so initiated, conduction persists until this voltage is reduced to a small value. This behaviour is suggestive of a gaseous diode with a very large voltage hysteresis, that is, one in which ionization and deionization voltages differ considerably. Thus, a simple relaxation oscillator employing a unijunction transistor generates a high-amplitude sawtooth voltage wave across the charging capacitor. In the circuit of Fig. 4.23, capacitor C is charged through resistance R_1 from the power supply. A small current flows between B_1 and B_2 and serves the function of back biasing the pn diode that effectively exists between E and B_1. As a consequence, negligible conduction prevails between E and B_1. However, when the voltage across C attains a sufficiently high value, the distribution of potential gradients within the device is suddenly changed. Specifically, the E–B_1 diode section becomes *forward* biased and therefore highly conductive. The capacitor is then rapidly discharged to a voltage low enough to return the E–B_1 diode to a reverse-biased condition. It then appears as an open switch and the cycle is repeated as the capacitor again charges toward firing potential.

It is seen that in addition to sawtooth pulses, short-duration spikes are also provided by this simple circuit. These spikes, accessible at B_1 and B_2, are on the order of several tens of microseconds wide.

So-called 'multivibrator' circuits are shown in Fig. 4.24. These circuits

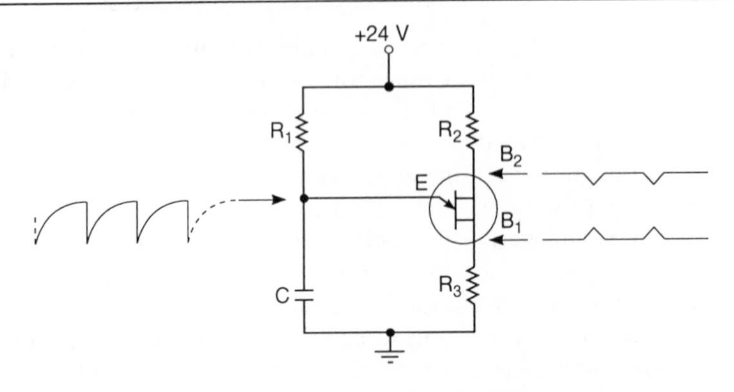

Fig. 4.23 *Unijunction transistor relaxation oscillator*

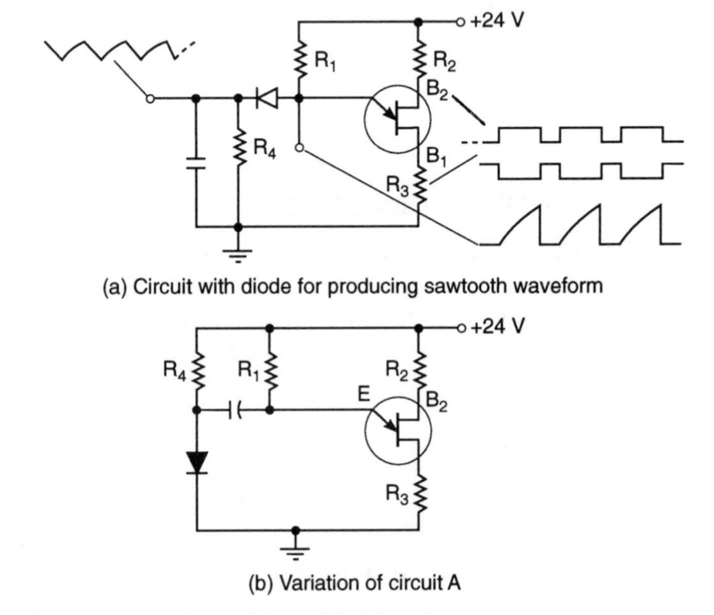

Fig. 4.24 *Unijunction transistor multivibrator circuits*

provide waveforms similar to those obtainable from two-element oscillators of the Eccles–Jordan type. The essential difference between the circuits of Fig. 4.24 and the relaxation oscillator of Fig. 4.23 is the inclusion of a diode to prevent discharge of the capacitor through the emitter-Base-1 circuit of the unijunction transistor. Rather, the capacitor must discharge through resistance R_4. This requires appreciable time; during this time, the voltage at the emitter relative to Base 1 is sufficient to maintain conduction. When the capacitor is nearly discharged, the voltage at the emitter falls the tiny

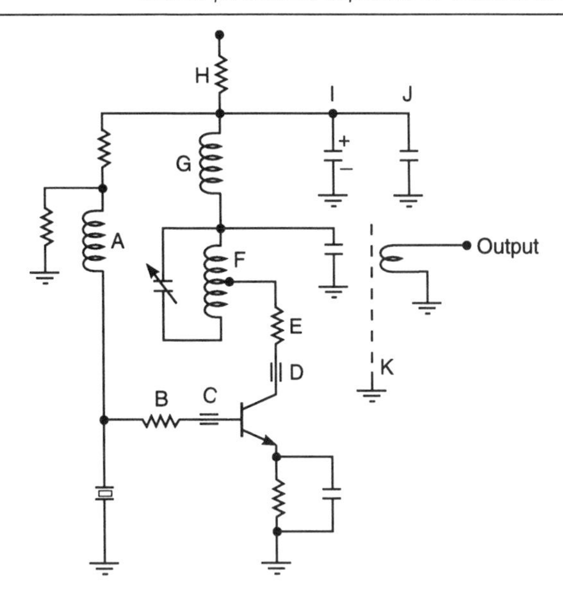

Fig. 4.25 *Band-aids and improvement techniques for a bipolar Miller crystal oscillator. A, radio frequency choke in bias circuit makes for easier starting. B, C, D, E, Resistors and/or ferrite beads discouraged high frequency parasitics. F Tapped tank inductor can optimize impedance match to transistor. G, Collector radio frequency choke should be a different value from base-circuit choke A. H, I, J, Decoupling network discourages low-frequency parasites. K, Fraday shield helps isolate oscillator from load*

additional amount necessary to extinguish conduction. A new charging cycle then commences. We see that the diode performs as a hold-off device, enabling the capacitor voltage to trigger conduction, but preventing current from the charged capacitor through the unijunction transistor.

Optimizing the performance of the Miller crystal oscillator

The Miller-type crystal oscillator shown in Fig. 4.25 is a fantasy in the sense that all of the 'add-ons' are not likely to be encountered in a practical circuit. It serves the purpose, however, of illustrating some techniques available for improving the performance of oscillators that may malfunction in some respect, or which do not operate optimally. As shown, such a configuration would be useful for parallel-mode crystals from 100 kHz to 10 or 15 MHz.

The radio frequency choke designated by 'A' isolates the crystal from the shunting effect of the base-biasing resistors. This is useful in bringing about reliable starting behaviour, especially where the crystal may have marginal activity, the beta of the transistor is low and the biasing resistors are of relatively low value.

The resistors and ferrite beads called out by B, C, D and E are sometimes needed to suppress high-frequency parasitic oscillation. Ordinarily, only one such item is used and it is most likely to be a ferrite bead in the base lead. A resistor, when used, is often in the 10- to 22-Ω range. Ferrite beads are available with different permeabilities and with different dissipative-loss characteristics so a little experimentation may prove rewarding in discouraging a stubborn parasitic. Basically, the bead and the section of the conductor passing through it form a tiny radio frequency choke. Parasitic oscillations at VHF and UHF frequencies result from stray inductances and stray capacitances and are, accordingly, much dependent upon the constructional practices employed in the oscillator. Also, some transistors (often those with good high-frequency capability) are prone to this type of malfunction.

The tapped tank-coil shown by F is an impedance-matching technique that can dramatically improve oscillator operation, especially when outputs exceeding about 1 W are involved. An optimum tap improves the efficiency of both the resonant tank and the transistor. At higher power levels the collector circuit of the transistor tends to load down the parallel-tuned tank. At 'flea-power' levels, such a tap is generally not necessary. Also, other things being equal the tap is more needed for transistor oscillators that operate at low collector voltage and high collector current than for the converse d.c. operating format.

Attention is directed to the usually-used radio frequency choke, G, to point out that its inductance should be different from any other that might be incorporated in the oscillator, such as the choke indicated by A. (Another example of a second radio frequency choke would be found in Colpitts oscillators where a choke is often used in the emitter circuit.) Identically-valued chokes in two parts of an oscillator circuit often cause mysterious oscillation at a lower than intended frequency.

Another source of low-frequency 'parasitic' oscillation in bipolar transistor oscillators comes about from the manner in which such transistors are operated at high frequencies. Intended oscillation usually occurs quite far down on the current-gain characteristic of the transistor. Although the current-gain may be more than sufficient for the intended oscillation mode, the fact remains that much-greater gain is available at lower frequencies. Therefore, if the transistor 'sees' any opportunity to oscillate at a low frequency, it will do so. The two possibilities ordinarily encountered are the previously-alluded to 'TPTG' oscillations provoked by radio frequency chokes, and relaxation oscillations made possible by appropriate time-constants in the power supply. A tightly-regulated power supply generally inhibits such relaxation-type low-frequency oscillation. Paradoxically, in some cases, it is necessary to sacrifice d.c. voltage regulation in a special way in order to discourage this type of malfunction. H, I and J form a low-frequency decoupling network that often proves effective. At low power-levels, say below 1 W, H may range from several-tens of ohms to several hundred of ohms. I is an electrolytic capacitor, usually in the vicinity of several hundred microfarads and is 'bypassed' at radio frequency by J, which can be a 0.01 μF ceramic capacitor. Finally, in order to reduce low-frequency gain, the emitter-resistor bypass capacitor should be no larger than that which suffices for the intended oscillation frequency.

The Faraday-shield designated by K is useful where output is derived via inductive coupling. Essentially a grounded electrostatic partition, this technique isolates the oscillator from the capacitive influences of the load. That is, the capacitance between the tank and output coil is no longer effective in reflecting load variations back to the oscillator. Although shown with a crystal oscillator, this technique is especially rewarding when used with self-excited oscillators or VFOs.

Although not shown, a small capacitor connected from collector to base is sometimes needed in the Miller oscillator circuit. This is particularly true for 100 kHz and relatively low-frequency oscillators. Above 1 MHz, the internal capacitance of the transistor itself usually suffices for the needed feedback.

In JFET and MOSFET oscillators, choke A is not needed and a ferrite-bead in the gate lead is common practice for parasitic suppression. Also, low-frequency parasitics is not a problem with these devices; accordingly, the decoupling network, H, I, J, has no relevancy.

Optimizing the performance of the Colpitts crystal oscillator

The bipolar transistor and FET Colpitts crystal oscillator are often quickly thrown together and readily yield satisfactory results for many purposes. More often than not, the capacitor divider consists of two equal-valued capacitors. The arrangement tends to work well for a combination of reasons. First, of course, is the relatively high stability provided by a crystal even though the effects of temperature and loading of the active device are ignored. Moreover, the Colpitts circuit is not only an inordinately simple one, but it possesses the inherent feature that harmonics are attenuated in the capacitor divider. Thus, wave purity is easily attained.

Notwithstanding, there are a number of things that can be done to optimize frequency stability. The more important of these can be tabulated as follows.

- Select a good crystal. The 'goodness' of a crystal is not altogether governed by its activity or willingness to oscillate. Its temperature coefficient is very important. If the general nature of the temperature coefficient of frequency change is known, it is possible to implement compensatory measures. A crystal with a relatively flat temperature coefficient over the anticipated ambient or operating range is particularly desirable. All things considered, AT-cut crystals operating in the vinicity of 4 MHz tend to have a useful combination of desirable features.

- The use of ordinary-sized, equal-valued capacitors in the capacitive divider is not the best way to isolate the crystal from the active device. Bipolar transistors, especially, have input impedances that vary greatly with temperature and with d.c. operating current. It therefore makes more sense to make the base-emitter capacitor of the divider three to perhaps five times larger than its mate. In this fashion, the input impedance of the transistor is considerably masked and prevented from unduly influencing the crystal frequency. In practice, the same objective can be accomplished by making both capacitors large and equal. The general idea is to select sizes that have a reactance of about 50 Ω at the frequency of oscillation. This is largely an empirical matter and has to do with the willingness of the circuit to oscillate with very large capacitors—much depends upon the Q of the crystal, the beta of the transistor and the circuit impedance from emitter to ground.
- Use a buffer amplifier to isolate the oscillator circuit from the driven load.
- Often the series-resonant (Clapp) version of the oscillator contributes toward stability because the crystal then presents a relatively low impedance across its 'tank' terminals. This makes this resonance-mode less susceptible to the active device than is the case when the crystal is operated in its parallel resonant mode.
- Use a regulated source of d.c. operating voltage. Bipolar transistors, especially, undergo wide impedance variations with changing d.c. voltage.
- Use good wiring and fabrication techniques and make certain that there is high physical and mechanical integrity. These matters are, unfortunately, often overlooked because it tends to be assumed that the crystal and solid-state device take care of everything. However, when one considers the several picofarads of trimming capacitance that often suffices to centre the crystal frequency, it is apparent that small changes in the circuit strays cannot be neglected. Not only is stray capacitance change likely to cause a frequency deviation, but it does not take too many nanohenries of inductance to do the same. This argues for short leads rigidly secured. Naturally, the best approach is high-quality printed circuitry.

Figure 4.26 shows practical implementation of these suggestions. It incorporates the above-suggested circuitry approaches. Another useful technique

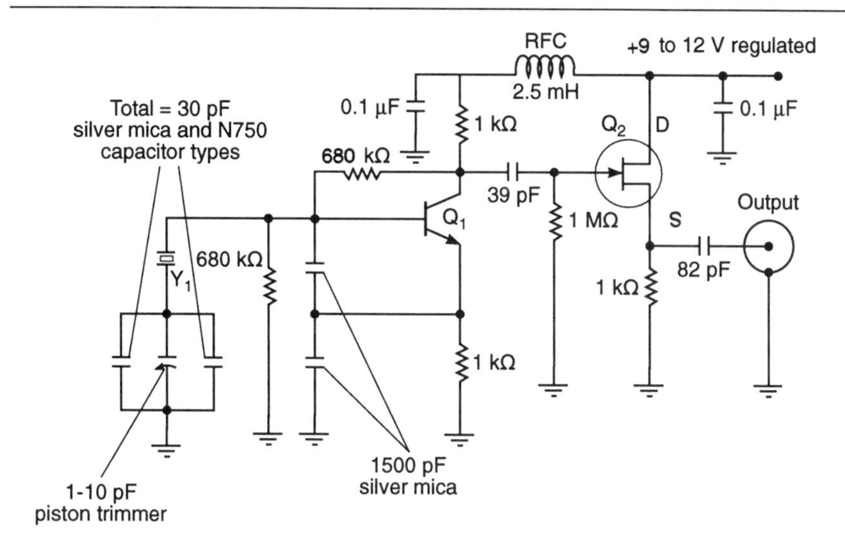

Fig. 4.26 *Oscillator using special techniques to optimize frequency stability. A 4 MHz AT-cut crystal is series-resonated in a Clapp oscillator circuit which employs heavy capacitive swamping of the transistor input impedance. Oscillator output is derived from a frequency insensitive point and capacitively coupled to a Class-A FET buffer. A voltage-regulated power supply is used. Crystal and circuit-temperature coefficient is empirically brought to a low value via optimum combination of silver mica and N750 crystal load capacitors*

might be a constant-temperature oven for the crystal. This suggestion was kept for the last because the crystal oven is too often held to be the panacea for frequency deviation. Its use can dramatically improve stability, but only after the various other influences on frequency constancy are given due consideration. In this regard, ovens using proportionate control rather than on–off temperature cycling are to be preferred. The output of the circuit shown in Fig. 4.26 can be subjected to subsequent power-boosting or frequency multiplication without adversely affecting frequency stability.

5 Universal oscillator circuits

For the most part, the oscillator circuits discussed in the previous chapters are readily implementable as practical oscillators utilizing one or another active device. This, indeed, was the objective of using the concept of the *universal amplifier* in the basic oscillator circuits. Insight into the nature of practical oscillators can be even further enhanced by scrutiny of a number of oscillator circuits utilizing specific active-devices and designated component-values. Accordingly, this chapter will depict and discuss numerous practical oscillators. These will be configured around the various solid-state devices and ICs that are now available and are represented by the symbol of the universal amplifier.

The universal amplifier: a three-terminal active device

It would be well to reiterate a few facts pertaining to oscillators and their active devices. Most of the active devices that will be found associated with the oscillator circuits are three-terminal amplifiers. Of course, there are negative-resistance diodes and there are devices with more than three terminals, such as the dual-gate MOSFET. However, these stand out as exceptions. (Even with these devices, the resonant tank in the oscillator 'sees' exactly the same kind of negative resistance as is conferred via external feedback paths. Whether one states that an oscillator operates because of positive feedback or because of negative resistance, the two processes are mathematically similar. Also, it is of interest that devices with more than three terminals can be shown to depend upon three 'main' terminals or electrodes for their amplifying and oscillating action.)

In actual practice, it is convenient to deal with the fact that most solid-state devices function as three-terminal amplifiers and a few two-terminal devices (diodes) behave as if they already have internal positive feedback paths. Of interest, too, is the fact that circuits can be devised in which any of the terminals of three-terminal devices function as the 'common' terminal. Thus, while it is true that the source of a JFET is the 'intended' common terminal, either the gate or the drain can be appropriately connected to serve as the common (or grounded) terminal in the oscillator circuit.

It should also be recalled that the active device in an oscillator must provide power gain. It is sometimes erroneously thought that voltage or current gain will suffice. If this were true, an ordinary transformer would be able to provoke and sustain oscillation in a resonant circuit or its equivalent. This power requisite is predicted on fundamentals and applies to negative resistance devices and switching devices, as well as 'linear' amplifiers.

100 kHz transistor Butler oscillator

Although better results are now attainable via higher frequency crystals, the 100 kHz crystal has long been used for precise calibration, timing and instrumentation purposes. Although most of these crystals were intended for tube oscillators, it is usually possible to obtain equally good if not better stability with transistors. This is because transistors are relatively free from ageing effects and contribute almost negligible heat to the crystal. For many practical applications, the transistor version of the 100 kHz oscillator can dispense with the crystal oven often used to stabilize against temperature variation. Although transistors themselves are subject to thermal effects, the Butler circuit of Fig. 5.1 is minimally affected by small variations about room temperature.

In the Butler oscillator, the crystal oscillates in its series-resonant mode. Usually, these 100 kHz crystals can be pulled into exact frequency through the use of a small series variable capacitor, as shown. Q_2 provides a considerable amount of buffering action so that the load does not have as much influence on the oscillation frequency as is often the case with single-transistor oscillators. Notwithstanding, it is generally a good idea to experiment with the output coupling capacitor, making it as small as possible for the particular application. Inasmuch as the transistors are d.c. coupled in this Butler circuit, the wide tolerances of transistor betas ordinarily encountered may require some experimentation with the d.c. bias on Q_2. This can conveniently be brought about by shunting the 2.2 kΩ collector resistor of Q_1 with resistance several to many times this value, or by introducing a resistance path from the base of Q_2 to ground. The two procedures will produce opposite effects on the base-emitter bias of Q_2. The objective of such experimentation is not merely to get operation right for reliable oscillation, but to minimize harmonic production.

From the above, it would appear that harmonic generation, where

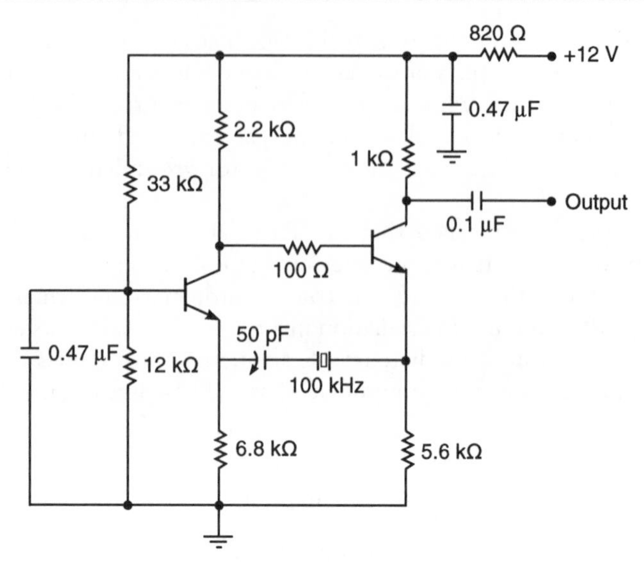

Fig. 5.1 *100 kHz Butler oscillator using npn bipolar transistors. Series-mode resonance of the 100 kHz crystal is tunable for calibration purposes*

desired, is readily attainable by biasing Q_2 in its nonlinear region. This, however, is detrimental to frequency stability and is likely to adversely affect start-up of the oscillator. It is much better to use subsequent squaring stages to produce harmonic energy. Conversely, the Butler circuit is in its element when it is desired to obtain very good wave purity.

An example of a dual-gate MOSFET oscillator

Figure 5.2 shows three Pierce oscillators utilizing dual-gate MOSFETs. The application is for beat-frequency oscillators enabling the operator of a communications receiver to receive either the upper or lower sideband of a single-sideband signal, or to copy code (cw) messages by means of a beat note. This triple-oscillator circuit is intended for use in conjunction with a 455 kHz intermediate frequency (IF) channel. In many instances, a single buffer amplifier will probably facilitate implementation.

Note that the two gates of the MOSFETs are connected together. This is the simplest way to use these devices in oscillator circuits and, for practical purposes, yields results comparable to similar circuits using JFETs. Inasmuch as the MOSFET is an inordinately good high-frequency amplifier, the same circuit configuration can be used for higher frequency IF channels. However, it would then be expedient to reduce the size of the gate capacitors and

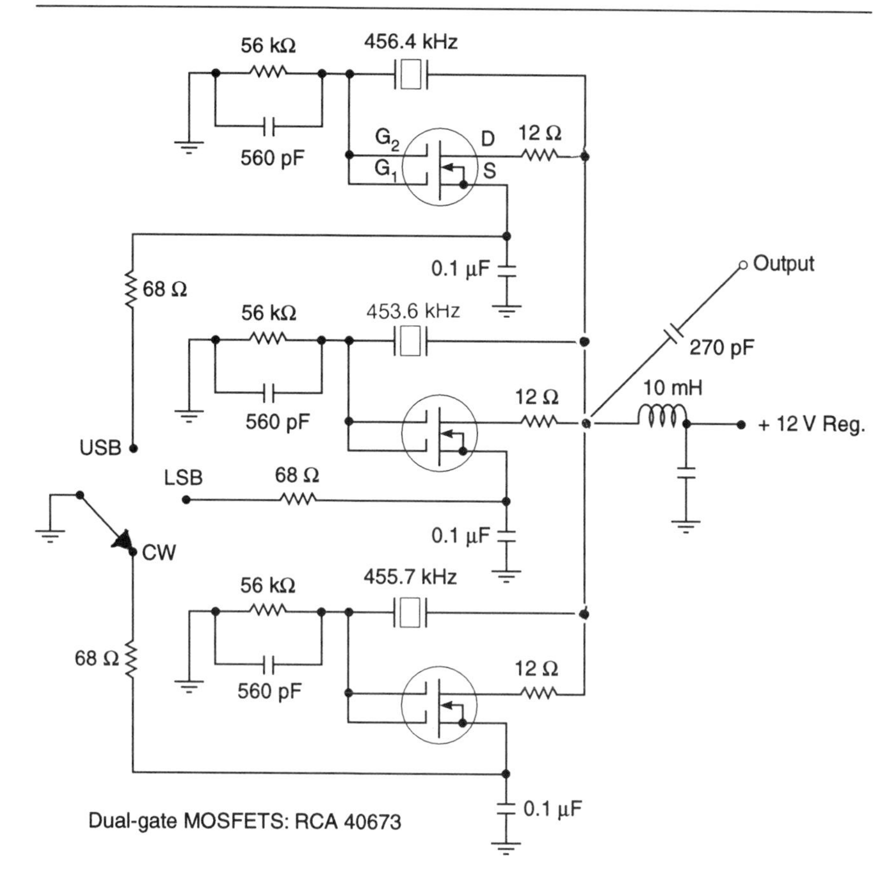

Fig. 5.2 *Example of dual-gate MOSFETS in triple BFO arrangement. Each oscillator is a Pierce circuit. By connecting the two gates together, the operation resembles that of JFETs*

of the radio frequency choke. For example, at 10 MHz the gate capacitors can be in the vicinity of 30 pF; similarly, the radio frequency choke can be a 500 μH unit. (Whether or not a buffer is used, the output coupling capacitor would likely be several tens of pF or less.) Other than these changes, it would be necessary to order 10 MHz crystals with the exact frequencies having the same arithmetic displacement from centre IF as the 455 kHz crystals.

The 12 Ω resistors in the drain leads of the MOSFETs discourages VHF and UHF parasitics. An appropriate ferrite bead often suffices for this purpose and in stubborn cases both the resistance and the bead can be used together. Sometimes these suppression techniques are implemented in the gate circuit. In any event, good high-frequency wiring and fabrication techniques should be used. On the other hand, MOSFET and JFET devices are not likely to produce the low-frequency 'parasitics' that often plague oscillators using bipolar transistors. (The gain of bipolar transistors is often much higher at audio frequencies than at the operating frequency of the oscillator. This is not true in most MOSFET and JFET oscillator applications.)

At first thought it might appear more straightforward to employ a single active device in which switched crystals provide the three frequency functions. Although this is sometimes done, the precautions necessary (especially for the higher IF channels) to isolate 'hot' crystal terminals from stray capacitances render the single-device design somewhat less than simple to implement. And whereas three active devices would have been awkward and uneconomical with electron tubes, this is certainly not the case in solid-state systems.

Single transistor parallel-T oscillator

The parallel-T (or bridged-T or twin-T) oscillator shown in Fig. 5.3 uses a single general-purpose npn transistor, such as the 2N2222 or 2N297. This approach represents a simple way to obtain a good sine wave without resort to large inductors and capacitors. With the T-parameters shown, oscillation occurs at approximately 1 kHz. If other frequencies are desired, the capacitors in the T network should be changed. These capacitors are inversely proportional to oscillation frequency. Thus, if the circuit is to be made to oscillate at 500 Hz, the series-arm capacitors should be doubled to 0.006 μF and the shunt-arm capacitor should be doubled to 0.012 μF. The 120 kΩ series-arm elements of the T-network should not be changed because they, in addition to their timing functions, are also involved in the biasing of the transistor.

The 20 kΩ variable resistance in the shunt-arm of the T-network is adjusted to make the circuit oscillatory. For the 1 kHz circuit shown, this will occur in the 7–11 kΩ range. Note that this shunt-arm resistance would be 60 kΩ according to the parallel-T design equations generally found in texts and handbooks. That is, the shunt resistance is depicted as being one-half the value of the series resistances. However, such a 'classic' T will not produce oscillation in the transistor circuit because its amplitude-null is too deep for the gain of the transistor to overcome. As it is, it will be found that higher-beta transistors will be more likely to oscillate, and that some low-beta specimens of a lot will not perform at all. The purpose of the variable shunt-resistance is to find an adjustment where the amplitude null is not too deep and which will still provide the requisite 180° phase reversal. Coincidence of these requirements occurs over a narrow adjustment range and it will be found that not very much variation of oscillation frequency is feasible.

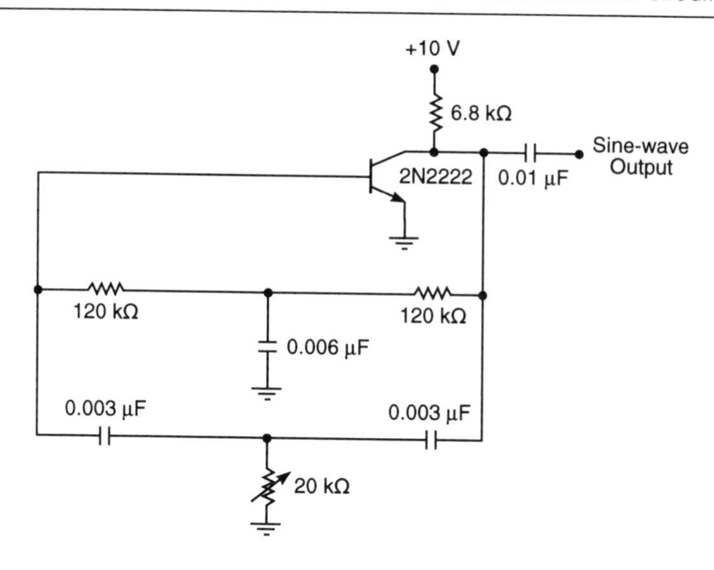

Fig. 5.3 *A single-transistor parallel-T oscillator. The natural harmonic discrimination of the T network results in a very low distortion sinusoidal output. The T values shown yield an oscillation frequency of approximately 1 kHz*

From what has been said, it can be correctly inferred that this is not a very vigorous oscillator and will not stand much loading. Therefore, an emitter-follower as a buffer amplifier may be found necessary for some applications. On the other hand, the T network is not readily pulled in frequency once its values are optimized for oscillation. Very high beta transistors may oscillate strongly at the expense of wave purity; this can be remedied by inserting a bias current-limiting resistance in series with the base lead. Such a resistance should be bypassed at audio frequencies—a 0.1 μF capacitor will usually suffice. Such an RC combination will have negligible effect on the oscillation frequency. For very low frequencies, consideration must be given to the electrolytic capacitors that may be used. Polarity should be observed and low-leakage types, such as solid-state tantalums, should be selected.

Several special-interest feedback circuits

Although a large number of oscillator circuits were explained in the previous chapters, these were by no means all inclusive. They were selected for discussion largely because of popularity of use, and also because of significant differences in configurations and/or mode of operation. Now that most small-signal oscillators are solid-state rather than electron-tube types, several unique oscillator circuits are often seen that would have been awkward to

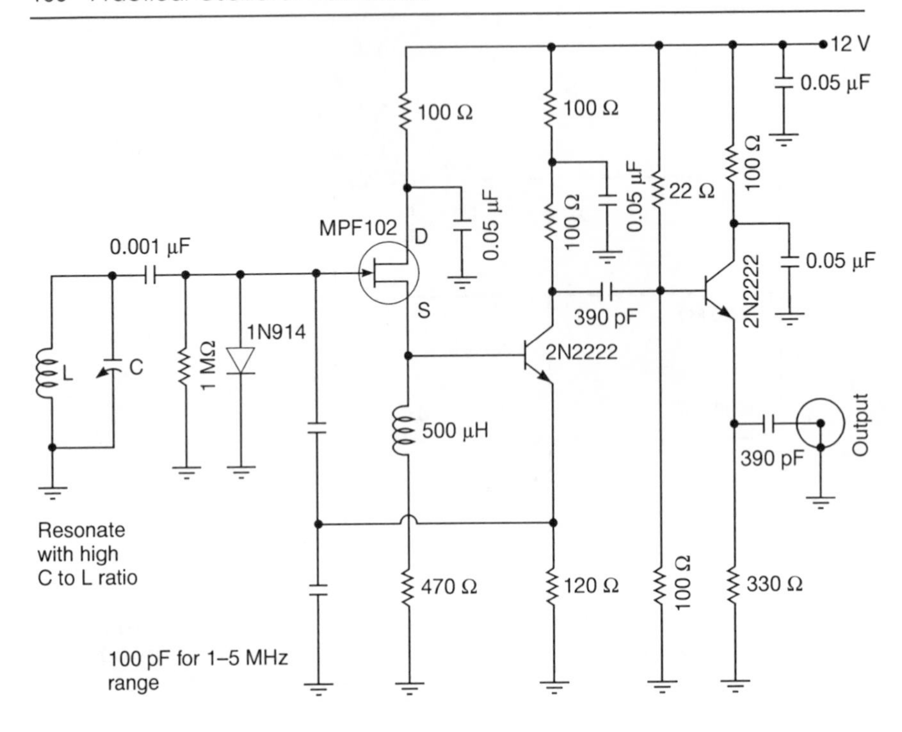

Fig. 5.4 *A typical Goral oscillator. The basic idea is to insert buffer amplifiers in both the feedback and the load path. This isolates the frequency determining portion of the oscillator from disturbances*

implement with tubes or did not merit much consideration for other reasons when tubes dominated the scene. Although 'new' oscillator circuits usually turn out to be derivatives of Colpitts, Hartleys or Miller circuits, it often happens that the modification, even if slight, contributes significantly to stability, or to some other performance feature.

Three such oscillators that have appeared in technical literature devoted to amateur radio are the Goral, Seiler and Vackar circuits. They all claim inordinate frequency stability for VFOs (self-excited oscillators using LC resonating circuits). In all three, the resonant circuit is protected in some manner from being degraded in Q, or from being detuned by the active device. This, of course, suggests the long-popular Clapp version of the Colpitts oscillator in which the series resonant tank presents a low-impedance to the input of the active device. However, the three previously-named oscillators use different techniques to accomplish essentially the same result.

The Goral oscillator of Fig. 5.4 has an emitter-follower stage inserted in the feedback path of an otherwise conventional Colpitts oscillator circuit.

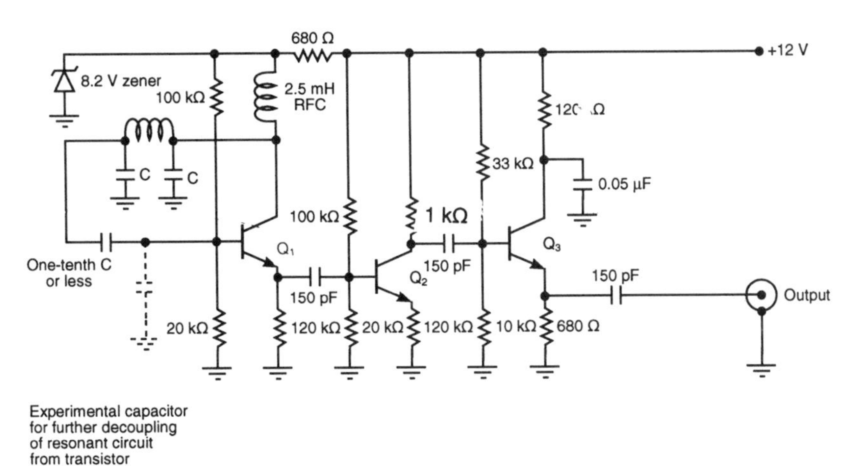

Q_1, Q_2 and Q_3 are 2N2222 or similar transistors. An MPF102 JFET can also be used for Q_1

Fig. 5.5 *Vackar oscillator*

The midpoint of the capacitive divider (which is actually part of the resonant tank) now sees a much lower impedance with respect to ground than would be the case without the emitter follower. The power gain of the JFET-bipolar transistor combination is much greater than that of the JFET 'oscillator' alone. There is latitude for considerable experimentation in the ratio of the two capacitors used in the Colpitts section of the circuit. This ratio can be optimized for frequency stability without easily running out of feedback. The 100 pF values shown are intended for a first try and will probably ensure oscillation.

Note that the emitter-follower is directly coupled to the JFET. It may be necessary to experiment with bias-determining resistances to ascertain Class-A operation from the emitter-follower. (Also, the output transistor is intended to operate in its Class-A region.) With commonly-used resonant LC combinations, this circuit can perform exceptionally well in the 1–5 MHz region. Those versed in high-frequency techniques will find little difficulty in extending this range. In general, it will pay dividends in frequency stability to strive for a high Q resonant tank via a high ratio of capacitance to inductance.

The Vackar oscillator circuit incorporates a π-section tank to attain the needed 180° phase-reversal in the feedback loop. However, the inverted feedback signal is not directly fed back to the input of the active device; rather, it is loosely coupled through a very small capacitor. Often, a shunt capacitor is introduced to further reduce the coupling. The basic idea is to isolate the resonant circuit as much as possible from the input of the active device, consistent with obtaining reliable oscillation. An example of a Vackar oscillator is shown in Fig. 5.5. Two Class-A buffer amplifiers are used for load isolation. Note the possible use of the shunt capacitor in dotted lines.

This circuit is particularly advantageous with solid-state devices, and especially with bipolar transistors that have inordinately-low input impedances and that present a widely-varying reactance to the tuned circuit as a consequence of temperature and voltage changes. Tuning can be accomplished by means of a small variable capacitor in parallel with either shunt-arm (C) of the π network. Although equal values of C are shown, once the overall circuit is operational, these values may be optimized for best stability. Generally, it will be found that the capacitor closest to the base of the transistor can be several times larger than the capacitor associated with the collector circuit.

It is interesting to note that if the small decoupling capacitor in the feedback loop were shorted out, the operation would be essentially that of the Pierce oscillator, noted for its vigorous feedback and its propensity for harmonic generation, but *not* for frequency stability without the use of a crystal. The introduction of attenuation in the feedback loop (via the small capacitor in the Vackar) prevents over-excitation and effectively isolates the resonant circuit from the active device. In further pursuit of this goal, the output is derived from the emitter, rather than the collector circuit. In this oscillator, it is important that the radio frequency (choke in the collector circuit 'looks good' at the operating frequency (presents a high inductive reactance). Resonances from distributed capacitance in the choke windings, especially those in the series-resonant mode, can degrade stability or even inhibit oscillation. Ferrite-core chokes are generally suitable for this application. (Sensitivity to radio frequency choke characteristics is common to all oscillator circuits that use chokes for shunt-feeding the d.c. operating voltage to the oscillator.) Unlike the Clapp, Miller and the conventional Colpitts oscillators, the Vackar does not readily lend itself to the substitution of a crystal for the LC tank circuit.

The Seiler oscillator has close topological resemblance to a parallel-tuned Colpitts. A basic difference, however, is the location of the d.c. blocking capacitor conventionally inserted in the base, drain or grid circuit. In the ordinary Colpitts, this capacitor is electrically close to the active device. In the Seiler version of the circuit, this capacitor is electrically close to the parallel-tuned tank. Moreover, in the Seiler oscillator this capacitor is made no larger than is necessary to reliably start and sustain oscillation.

The practical consequence of this slightly-modified configuration is that greater isolation of the resonant circuit is attained than in the conventional Colpitts. It can be argued that arbitrarily high tank circuit isolation can be achieved in the conventional Colpitts by appropriate manipulation of the

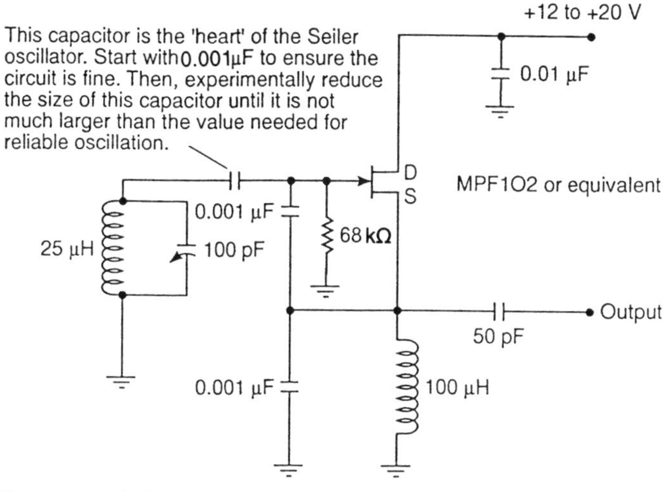

Fig. 5.6 *Example of a practical Seiler oscillator. The potentiality for excellent frequency stability inherent in this circuit is best approached by using a high C to L ratio in the resonant tank and by the use of a buffer amplifier and a regulated d.c. source*

size and ratio of the capacitors in the feedback divider network. However, in the Seiler version of the circuit, it will be noted that the parallel tank is isolated not only from the active device, but also from the capacitive divider in the feedback network.

Figure 5.6 shows a practical Seiler oscillator circuit. In order to focus attention on the oscillator itself, a buffer amplifier is not shown. However, as with all oscillator circuits from which optimum stability is desired, it will be found profitable to insert a Class-A buffer amplifier between the oscillator output and the load. In this particular oscillator circuit, a JFET source-follower would provide nearly ideal load isolation. Then, if more output voltage or power is needed, a second stage using a bipolar transistor as a common-emitter amplifier would suit the needs of many applications.

It will be noted that the 1N914 or similar diode usually shown in the gate circuit of JFET oscillators is not present in the Seiler oscillator of Fig. 5.6. This is because of the low degree of feedback that results from deliberately making the tank-isolating capacitor barely large enough to maintain oscillation. Under such an operating condition, the pn input diode of the JFET will not be driven into forward conduction and therefore needs no external diode for protection. Because of crystal-holder capacitance, it does not appear that the isolation technique used in the Seiler oscillator lends itself well for use with crystals.

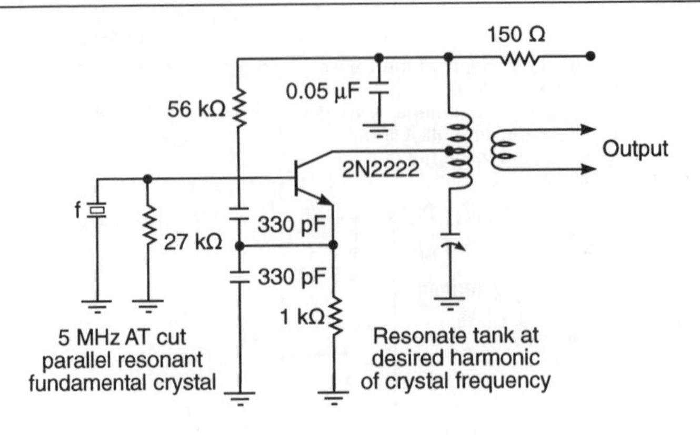

Fig. 5.7 *Example of harmonic oscillator. In the above circuit, all frequency multiplication takes place within the transistor and is governed by the resonance of the output tank circuit. The position of the collector tap should be empirically optimized for highest output of the desired harmonic frequency*

A harmonic oscillator using a fundamental-frequency crystal

Oscillator circuits are very much like people—it is often the little differences that count. Whereas the circuit of Fig. 5.7 may superficially appear to be just another version of the various configurations already discussed, there is a significant difference in the way it operates. The circuit is basically that of the Colpitts with the apparently minor variation of a tuned-tank circuit in the output. Note that the inductor tap is not arranged in the appropriate way to make a Hartley oscillator, despite a superficial resemblance thereto. In this scheme, the sole purpose of the tap is to provide a better impedance match to the collector circuit of the transistor. This is more important here because the output frequency is at a harmonic of the crystal frequency.

The important aspect of this oscillator is that harmonic energy is delivered from the circuit although an overtone crystal is not used. Indeed, the crystal, a fundamental-frequency, parallel-resonant type, does not vibrate in overtone modes. Although this method of attaining crystal stabilization of higher frequencies is particularly efficient when tuned to operate as a frequency doubler, higher multiplications can be achieved by resonating the collector tank-circuit at the desired harmonic. A small tuning capacitor connected across the crystal will permit greater frequency-pulling than is readily obtained from an overtone crystal.

If one is starting from scratch in the design of a transmitter or receiver, this

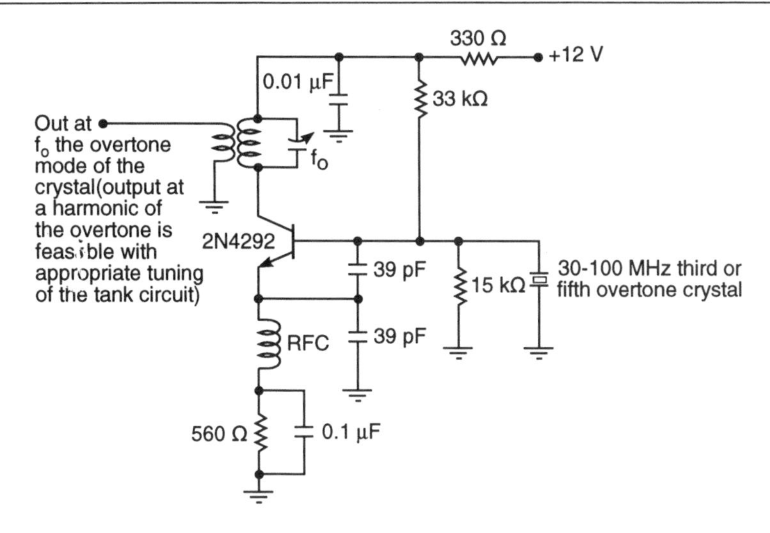

Fig. 5.8 *Overtone oscillator with provision to discourage fundamental frequency oscillation. The inductance of RFC must be large enough to permit operation at* f_o*, the overtone mode, but small enough to prevent oscillation at the fundamental frequency of the overtone crystal*

approach merits consideration for several reasons. Parallel-resonant crystals for fundamental frequency operation are more available than overtone types. Being AT-cut types, they have relatively good temperature stability. This may be true also of third-overtone crystals, but fifth, seventh and ninth overtone crystals are often BT cuts and tend to have poor temperature-coefficients. As mentioned, fundamental crystals are more amenable to frequency pulling than overtone types. Finally, fundamental crystals are generally less expensive than overtone crystals.

This, however, is not the end of the story. It is possible to combine the two techniques in order to reach a specified frequency. That is, many of the overtone-crystal oscillators can be used in conjunction with an output tank tuned to a desired harmonic. In such instances, the overtone-crystal oscillators at an overtone—usually the third, but further harmonic multiplication takes place within the active device and is selected by the tuned output tank. This can prove a bit tricky and is best attempted with a scheme such as depicted in Fig. 5.8. Here, the emitter inductor, RFC, encourages the crystal to operate in its third overtone mode. At the same time, it discourages vibration in the fundamental mode.

Usually, the oscillators that act also as frequency multipliers exhibit good buffering action with high immunity to frequency pulling from the driven circuit or load.

A bipolar transistor overtone crystal oscillator

An overtone crystal oscillator circuit using an npn transistor is shown in Fig. 5.8. In this type of oscillator, the crystal itself oscillates at an odd harmonic of its fundamental mechanical frequency. This is in contradiction to circuits in which the harmonic multiplication takes place within the active element and in which the crystal oscillates at its fundamental mechanical frequency. Indeed, it is often necessary to prevent fundamental-frequency oscillation of the crystal in overtone oscillator circuits. One way of accomplishing this is through the use of a radio frequency choke in the emitter lead, as is seen in Fig. 5.8. The idea is to have sufficient inductance to enable oscillation to occur at the desired harmonic, but insufficient inductance to support oscillation at the fundamental frequency. This generally requires experimentation and may be facilitated via the use of a slug-tuned inductor. The higher the order of the desired harmonic, the easier it is to determine a suitable inductance for RFC, providing one does not run into trouble with stray resonances.

The reason fundamental frequency oscillation is undesirable in the overtone oscillator is that this lower frequency will generally contaminate the output and will tend to ride through the entire receiver or transmitter system. Moreover, the simultaneous oscillation at the fundamental and desired harmonic is generally at the expense of harmonic output level. For many applications, overtone oscillators do not develop a surplus of power, at best. Accordingly, any technique that will result in more efficient production of the desired harmonic is worthy of consideration. Simultaneous oscillation at the fundamental and harmonic frequencies of the crystal is not always immediately obvious and can be a source of spurious frequencies within the system. In the technical literature, one often finds the Pierce oscillator used with overtone crystals because of the strong feedback developed at high frequencies by this oscillator circuit. This, however, can be a two-edged sword because of the readiness at which the crystal will vibrate in its fundamental frequency when used in this simple circuit.

Overtone crystals yield harmonics that are nearly, but not precisely integral multiples of the fundamental frequency. Moreover, they are not as readily pulled in frequency as are fundamental-frequency crystals. Their use is dictated by the physical and electrical fragility of fundamental-frequency crystals at frequencies higher than 50 or 60 MHz. Just where the demarcation is between fundamental and overtone crystals depends upon the state of the art, cost, the desire to dispense with frequency multipliers and other factors. Overtone crystals, in any event, find wide application beyond the 10 MHz region, but may also be used at lower frequencies. Keep in mind that overtone crystals are made in such a way as to enhance vibration at a particular odd-harmonic. (Conventional crystals can, but are not likely to perform well in the overtone mode. They are best used with frequency

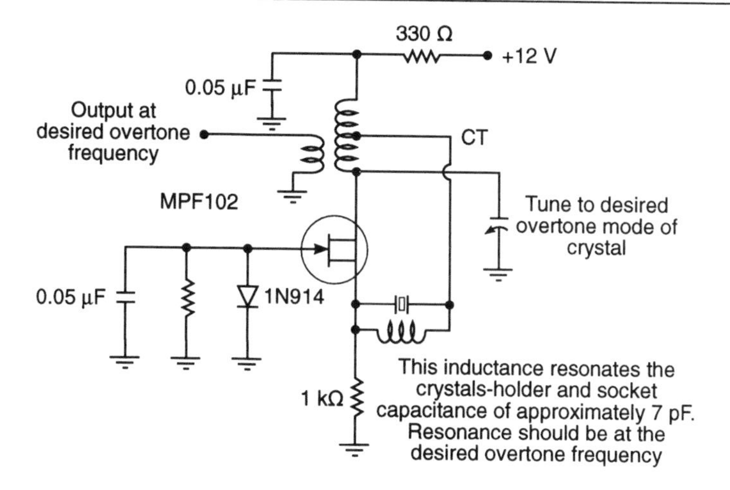

Fig. 5.9 *Overtone oscillator using a junction-field-effect transistor. This is a Hartley circuit and is dependent upon series-mode resonance of the crystal at a desired overtone frequency, tuned by the output tank circuit. Optimum efficiency, as well as suppression of fundamental-frequency oscillation, is brought about by resonating the crystal shunt capacitance*

multiplier stages or in oscillator circuits where frequency multiplication occurs within the active element.)

An overtone crystal oscillator circuit using a FET

Another scheme for exciting overtone modes in overtone crystals is shown in Fig. 5.9. This is not merely an FET substituted for the bipolar transistor of Fig. 5.8. There is more than meets the eye here. First, the Hartley oscillator circuit is not a trivial substitution for the usual Colpitts, for it happens that the Hartley tends to develop stronger feedback in many practical applications. This is favourable for excitation of the higher overtone modes of the crystal. At the same time, the gain of the FET does not progressively increase at lower frequencies as is usually the case with bipolar transistors operating in the radio frequency region. This makes it less likely that the fundamental mode of the overtone crystal will be excited.

A novel aspect of this circuit is the resonating inductance connected across the crystal. When this inductance has the same numerical reactance as the 7 pF or so of holder and socket capacitance peak performance in the desired overtone mode is attained. This further discourages any tendency of fundamental-mode oscillator. of course, some experimentation is required but this may be facilitated via the use of a slug-tuned inductor. This oscillator is particularly advantageous for operation at higher than the third overtone of the crystal. Fifth, seventh and ninth overtone modes can be generated and can be optimized by experimenting with the Q on the output tank and its feedback-tap. At the higher overtone-modes, best results will be experienced with light output-circuit loading. Accordingly, a buffer amplifier, preferably another JFET, will lead to best performance.

The use of diodes to select crystals electronically

It is often necessary to be able to switch-select different crystals in a crystal oscillator. This is easy enough to do when one crystal terminal connects to ground. For then, it is only a matter of wiring the 'hot' side of the crystals to the appropriate poles of a selector switch that has its rotor grounded. This technique, however, does not readily lend itself to situations where it is desired to select crystal frequencies from a remote point. Also, there are many oscillator circuits in which both crystal terminals operate at relatively high impedances above ground. Electronic switching neatly resolves the problems attending both of these situations. It also enables the use of a mundane switch rather than a special low-capacitance type.

Electronic switching is accomplished via the use of small-signal diodes. Figure 5.10 is an example of electronic switching of nominally 48 MHz crystals in a modified Pierce oscillator using overtone crystals. The Pierce is not only a good active oscillator circuit for exciting the overtone crystals, but it also provides the positive d.c. connection to the anodes of the switching diodes. The 2.4 kΩ resistances provide radio frequency isolation as well as limiting the bias current through the diodes. Thus, the switch can be some distance from the oscillator. The impedance of the forward biased diode is sufficiently low so as to impart negligible effect on the stability or activity of the oscillator. It is only necessary that the peak radio frequency voltage be much smaller than the bias voltage.

In the oscillator of Fig. 5.10, it will generally be necessary to retune the output tank circuit in order to optimize the overtone frequency of the selected crystal. In other situations, no adjustment beyond actuation of the crystal selector switch is required. For example, a Pierce oscillator can have diode-selectable fundamental-frequency crystals with no need for tuning. (Pierce oscillators using fundamental-frequency crystals often have a radio frequency choke in their output circuit rather than a tuned tank-circuit.)

Electronic tuning with a reverse-biased silicon diode

The effective capacitance evidenced across the terminals of a pn junction is a function of the impressed d.c. voltage. Unfortunately, the forward-based

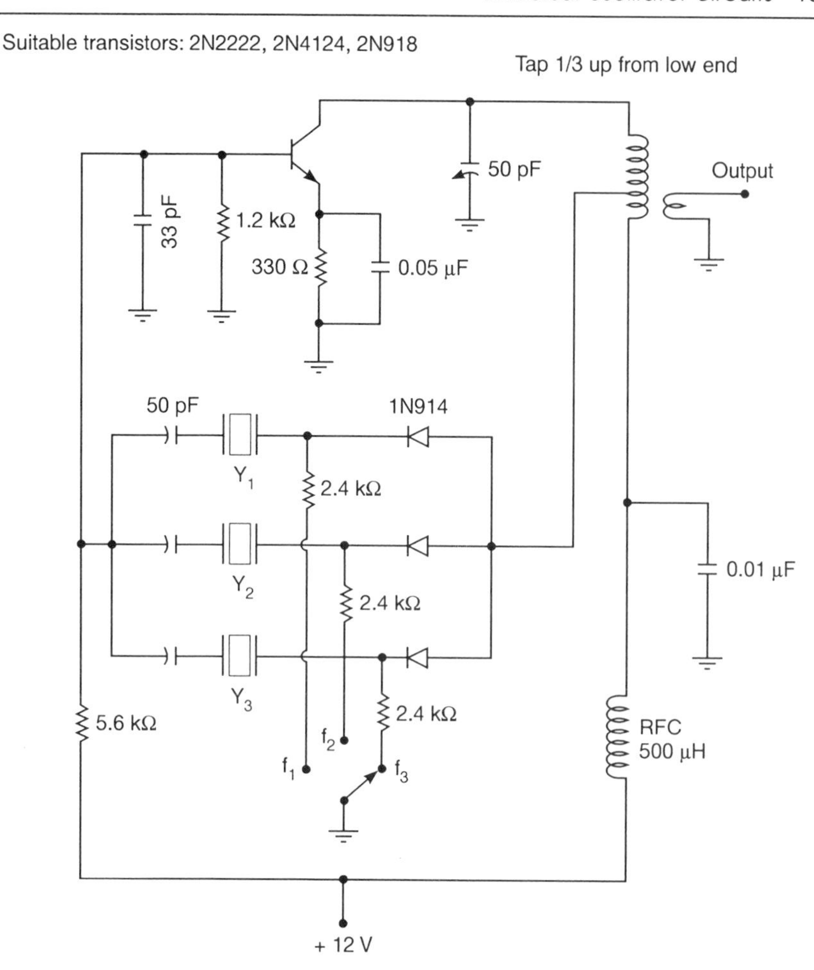

Fig. 5.10 *Example of diode switching of crystals. In the above circuit, the three overtone crystals are in the vicinity of 48 MHz. Selection of a frequency should be accompanied by resonating the output tank circuit*

situation is also accompanied by an effective dissipative resistance that precludes the use of the capacitance component as part of a resonant tank circuit. On the other hand, virtually no resistive dissipation is associated with the reverse-biased junction, thereby making it a suitable voltage-dependent variable capacitance for tuning purposes. Awareness of this basic feature of pn diodes readily leads to implementations for electronically-tuned oscillators, frequency/phase modulation applications, and remotely-ganged radio frequency amplifiers. A few practical aspects of this tuning technique should be kept in mind.

Figure 5.11 shows a generalized oscillator circuit for use in the 2–30 MHz

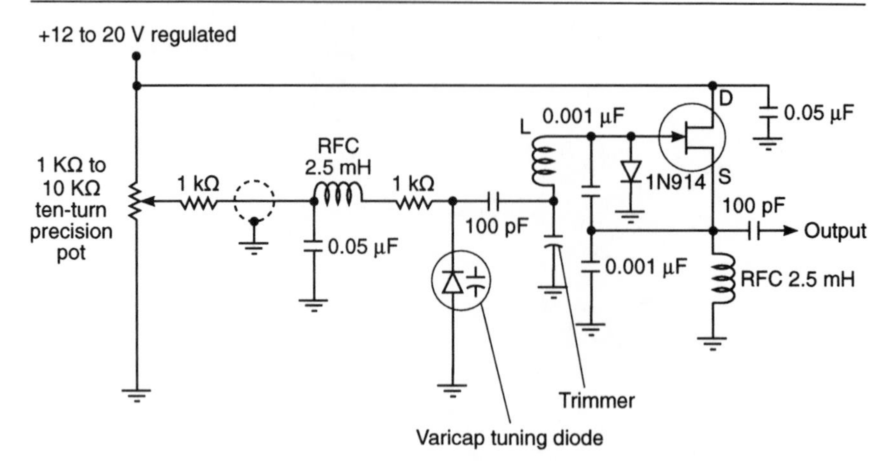

Fig. 5.11 *Example of varicap tuning of an oscillator. This arrangement shows the general features of varicap tuning for the amateur HF bands. Various varicaps are available for optimum results, but ordinary signal diodes often work well*

amateur bands. Often, either a specially manufactured varactor or 'varicap', or an ordinary pn junction diode will prove satisfactory. Much depends upon the required capacitance range, the demand made on uniformity of the diodes for production purposes, the shape of the capacitance versus voltage tuning curve and other factors. At higher frequencies a varicap will manifest a high Q than a common signal or rectifier diode.

Note that the tuning diode connects to the 'hot' junction of the Clapp series-resonant tank circuit in Fig. 5.11. This places a heavy burden on the radio frequency choke, the purpose of which is to isolate the d.c. control circuitry from the tank circuit. The 1.5 kΩ resistor helps relieve some of this burden, but it remains important that this radio frequency choke presents a high inductive reactance over the entire frequency range through which the oscillator is tuned. The control circuitry to the left of the 0.05 μF capacitor can be of indefinite physical length, but should be shielded to prevent EMI pick-up.

The best use of the scheme results from the use of a precision ten-turn potentiometer and digital read-out dial. A high-performance voltage-regulated power source is a must, for otherwise the tuning will not be stable and/or hum or noise modulation will be imparted to the oscillator. As is shown, it also makes sense to operate the oscillator from such a supply. Other techniques contributing to basic oscillator stability should also be implemented. For example, the Colpitts capacitive divider should utilize silver mica or polystyrene capacitors. Also, these capacitors should be large enough to effectively 'swamp-out' capacitance variations in the active device. (This is of greater importance with bipolar than with field-effect

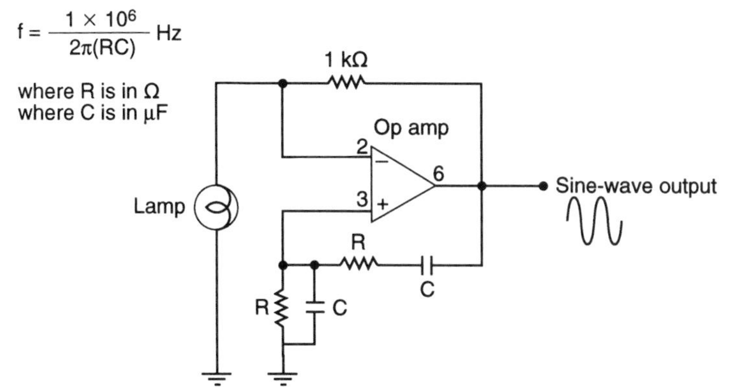

Op amp: General purpose such as LM741, LM747, LM748, etc.
Pinout may differ from circuit
Lamp: ELDEMA 1869 or General Electric 1869
Power supply: ± 10-15 V, with centre-tap to ground
For LM741, pin #4 is + Vcc, pin #7 is −Vcc

Fig. 5.12 *Wien bridge sine-wave oscillator using op amp. Positive feedback for oscillation is provided by the RC network. Variable negative-feedback for amplitude regulation is provided by the lamp filament*

transistors.) The inductor, L, should have high Q and excellent mechanical integrity. It may be found necessary, in some instances, to connect a several-thousand-ohm trimpot on the ground lead of the tuning potentiometer to prevent the radio frequency voltage from forward-biasing the tuning diode on negative excursions.

Wien bridge oscillator

Excellent sine-wave oscillators can be configured around various op amps. This is because op amps provide much greater gain than discrete devices and at the same time have significantly higher input impedances. This enables the use of various RC networks without serious loading problems. These oscillators are suitable for low-frequency work, where they dispense with the bulky inductors and capacitors needed for LC oscillators. General-purpose op amps will perform well up to the 100 kHz vicinity and op amps or voltage-comparators with higher frequency capability can extend this to several MHz, and beyond.

Figure 5.12 shows a practical implementation of the Wien bridge oscillator. In order to optimize the purity of the sinusoidal output wave, a little experimentation is in order so that the amplitude regulating action of the lamp in the negative feedback loop can be maximized. This can often be accomplished by experimenting with values above and below the 1 kΩ a resistance shown between the output and the inverting input terminal.

Sometimes, a few hundred ohms may be inserted in the ground lead of the lamp. It should be recognized that the primary purpose of the regulation is to minimize distortion. The amplitude stabilization attained by this technique is, of course, desirable, but is not the main goal in most applications of this oscillator circuit. The lamp automatically keeps the net positive feedback low enough so that the op amp works in its Class-A region, thereby avoiding the nonlinear amplitude excursions present in most oscillators.

Instead of lamp filaments, positive temperature coefficient thermistors can be used. These elements are available with smaller thermal time constants than lamps. This feature is important in ultra-low frequency oscillators where the resistance of a lamp filament will change during a single oscillation cycle. If this occurs, the distortion will be increased rather than reduced.

Tuning of this type of oscillator can be achieved by ganging either the R or the C elements together. Note that the R and C elements of the series arm can be transposed in order to accommodate a two-gang tuning capacitor with a common metal shaft. However, the rotor of the tuning capacitor must be insulated from ground.

The op amp square-wave oscillator

The basic configuration of a widely used multivibrator type oscillator is shown in Fig. 5.13. With the popular 741 op amp, high quality square waves can be generated up to 10 kHz or so. For higher frequencies, there are many op amps available with high switching speed capability. Even better results may be forthcoming from voltage comparators. Depending upon the internal circuitry of the op amp or voltage comparator, a couple of minor variations from the simple circuit of Fig. 5.13 may be necessary. In some instances, a load resistance of a kilohm or so may be needed between the output of the amplifier and the positive d.c. supply. Also, in some cases a high resistance, say on the order of 100 kΩ, should be connected between the amplifier's noninverting input (+) and the positive d.c. supply. (The addition of these two modifications to the circuit of Fig. 5.13 will not appreciably alter its performance and will thus impart universal application to the circuit.) The frequency can be changed by varying either the timing capacitor or the upper 100 kΩ resistance in the diagram. (The 100 kΩ resistance connected from the amplifier's output to its noninverting input provides the positive feedback path and generally should not be used for frequency manipulation.)

Although the 50% duty cycle square wave provided by the circuit of Fig. 5.13 is required in many applications, other applications are best carried out with other duty-cycle ratios. For such purposes, the basic oscillator of Fig. 5.13 can be readily modified to operate as an adjustable duty-cycle pulse generator. The manner in which this is accomplished is depicted in the

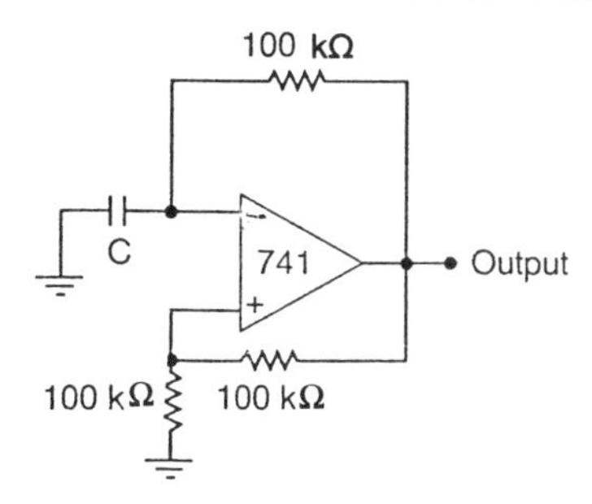

Fig. 5.13 *A square-wave oscillator suitable for a wide variety of op amps. A 50% duty cycle multivibrator type square-wave generation results from alternate turn on and turn off of the op amp*

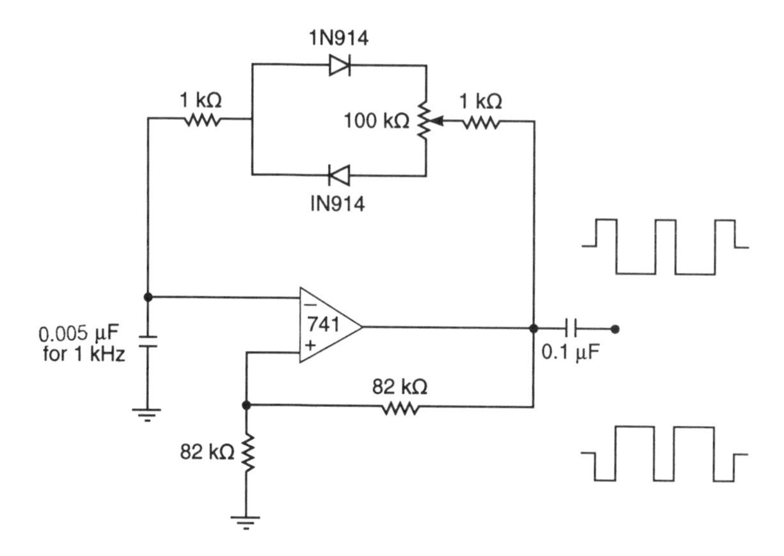

Fig. 5.14 *Adjustable duty cycle pulse generator. Output waveforms for extreme adjustments of duty-cycle potentiometer show that the frequency remains constant. This remains true even when the frequency is changed by changing the timing-capacitor, C*

circuit of Fig. 5.14. Here, diodes are used to steer the charge and discharge cycle of the timing capacitor through different resistive paths. The 100 kΩ potentiometer enables the adjustment of these resistive paths, and therefore the duty cycle. An interesting and useful feature of this circuit technique is that the pulse repetition rate remains substantially constant over the whole range of duty-cycle ratio. By means of a tapped switch, timing capacitors can be selected to provide useful repetition rates. Otherwise, the statements pertaining to the simpler oscillator of Fig. 5.13 also apply.

The oscillators of Figs 5.13 and 5.14 are capable of very low frequencies if special attention is given to the electrolytic capacitors which are then required for C. Low-leakage units are needed. Solid-state types are recommended. Frequencies down to and below 1 Hz are thereby feasible.

Oscillator using an IC timer

Although op amps contain many transistors and other simulated components, they are not always the best active device for implementing oscillators. There are a number of ICs which are, in essence, sub-systems and can be as superior to op amps as the op amp is to discrete devices. The kind of superior performance that accrues from these more complex ICs pertains to precision, stability and flexibility. For example, oscillators made with the popular 555 timer can be designed for a highly predictable frequency, have:

- high immunity to variations of power supply voltage
- internally implemented temperature stability
- better load isolation than is readily attainable through the use of buffer amplifiers.

The audio-oscillator shown in Fig. 5.15 is intended for hobbyist projects but is easily applicable to demanding instrumentation purposes. This is accomplished simply by paying heed to the stability and quality of the RC timing components. Many of the practical difficulties with LC resonant tanks are circumvented in this type of oscillator. It develops a square-wave output. If sine waves are needed, it is easy enough to insert an LC passive filter (either low-pass or band-pass) or an op amp active filter between the output (pin 3) and the load. This approach is a compelling one because the filter characteristics are far less critical than would be the LC parameters of a tank circuit in an 'ordinary' sine-wave oscillator. The formula for calculating frequency in this oscillator is:

$$F = \frac{1.44 \times 10^6}{(R_1 + 2R_2)(C)}$$

where F is the oscillation frequency in Hz, R_1 is as shown in Fig. 5.15 and is in ohms, R_2 is as shown in Fig. 5.15 and is in ohms, and C is as shown in Fig. 5.15 and is in microfarads.

For example, if $R_1 = 5000\,\Omega$, $R_2 = 100{,}000\,\Omega$, and $C = 0.01\,\mu$F, we have:

$$\begin{aligned} \mathrm{F} &= \frac{1.44 \times 10^6}{(5000 + 2 \times 100{,}000)(0.01)} \\ &= \frac{1.44 \times 10^6}{(205{,}000)(0.01)} \\ &= 704.2\,\text{Hz} \end{aligned}$$

Thus, the lowest adjustable frequency of the circuit of Fig. 5.15 is

(a)

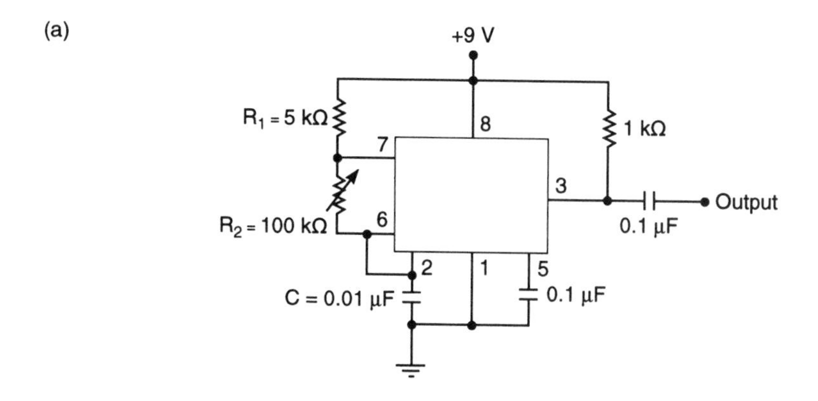

(b)

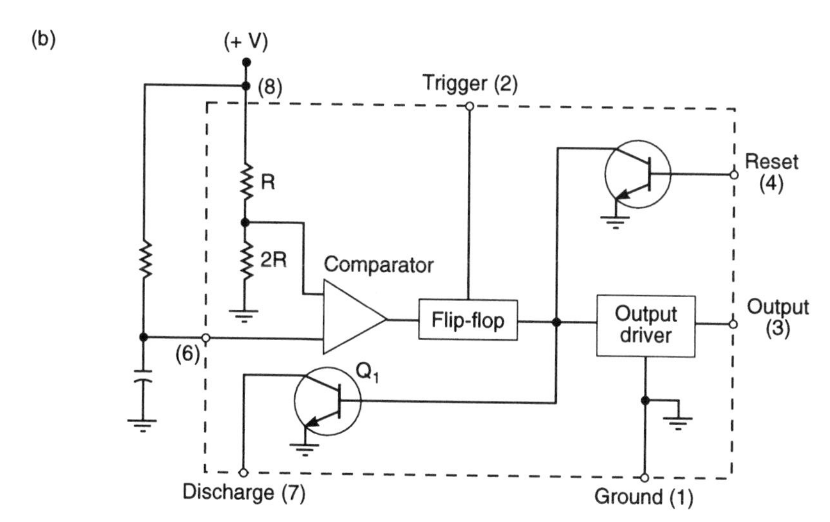

Fig. 5.15 *Audio oscillator using the NE555 timer IC. a, Circuit of oscillator suitable for code practice and as a general purpose source of audio frequency test voltage. b, Simplified block diagram of the internal circuitry of the NE555*

approximately 700 Hz. Higher frequencies are available by adjusting the potentiometer. This is a good frequency range for a code-practice oscillator. If headphones are used, however, this oscillator will drive a 4- or 8-Ω dynamic speaker if a 20 μF electrolytic capacitor is substituted for the 0.1Ω output capacitor shown.

A simple function generator

The function generator has been supplementing sine-wave test oscillators and signal generators and threatens to replace these two servicing and

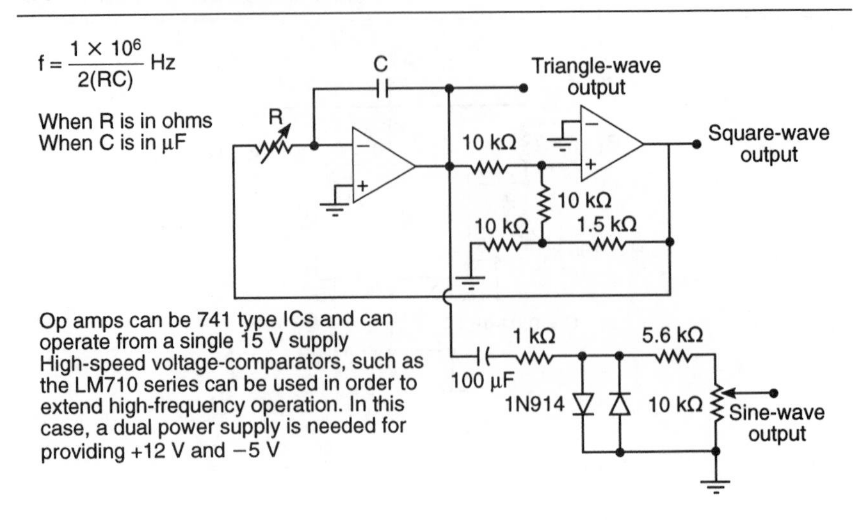

Fig. 5.16 *A basic function generator using general-purpose op amps. This type of oscillator provides optional outputs of square, triangular or sine waves*

calibration instruments. One reason is that the function generator makes available at least three waveshapes, sinusoidal, triangular and square. There are other reasons, too. The function generator readily allows various sophistications to be incorporated, and is particularly amenable to ultra-low frequency generation. On the other hand, function generators have been extending their high-frequency capabilities since they were first introduced. The result is that a single test instrument now can cover the frequency range previously requiring two instruments—the audio oscillator and the radio frequency signal generator.

Figure 5.16 shows a simple function generator. A frequency range of several Hz to several tens of kHz can be readily covered using inexpensive op amps. With selected op amps, especially with high-speed voltage-comparator types, several hundred kHz can be reached. Commercial function generators can perform out to several MHz and beyond because of special dedicated ICs. The circuit of Fig. 5.16 is 'bare-bones'; it enables the experimenter to construct this novel waveform generator with minimal complexity and cost.

The triangle to sine-wave converter is quite rudimentary, comprising the 1N914 diodes and associated resistors. A triangle-wave contains 15.5% of harmonic energy. These are odd harmonics—the third, fifth, etc. The two-diode sine-wave converter probably reduces this to the vicinity of 4 or 5%. This is, to be sure, a rather impure sine wave. It is, however, sufficiently good for a large variety of test and measurement purposes.

A more serious drawback is the high output impedance of the triangle to sine-wave converter. The impedances often encountered in solid-state circuitry would tend to load down the amplitude of the sine-wave output. An easy solution is to provide a voltage-follower or other type of buffer amplifier so that the sine-wave output will look like a low-impedance source. This was not included in the basic circuit of the function generator because the intent was to focus on the essentials comprising this interesting 'oscillator'. (In the interest of preventing the load from disturbing the frequency, it may be advisable to provide similar output amplifiers for the triangular and square-wave functions.)

The most convenient method of covering a wide range of generated frequencies is to use a multi-pole switch to select appropriate values of C. R provides the tuning and could be a precision tenturn potentiometer.

Square-wave oscillator using logic circuits

In the ensuing discussion, many oscillators using logic circuits will be dealt with. Like op amps, logic circuits are usually IC modules. Like op amps, logic circuits generally are comprised of many transistors. They even resemble op amps in that many of them have two inputs. (Some, however, have a multiplicity of inputs or gates.) Being digital devices, logic circuits are intended to be in either an on or off state so that an oscillator designed around such devices actually makes use of rapid switching action between these two states. Thus, we are not dealing with 'amplifiers' in the usual sense. This may appear somewhat puzzling at first, but it should be recalled that similar behaviour has been encountered with the op amp square-wave oscillators already described. The real difference in practical terms is that the op amp has a wide linear region between its on and off states. Conversely, the logic circuit has a very narrow 'linear' range between its two conductive states. The fact that both circuitries do have a linear range of amplification makes them qualitatively similar square-wave oscillators. The narrow linear range in logic circuits is compensated for by the extremely high gain available within the range.

Thus, from the standpoint of designing oscillator circuits, the logic device operates as a switching amplifier. This being the case, techniques for producing oscillation will still make use of phase inversion, feedback paths and the cascading of devices. Although an oversimplification, some insight is provided by considering the logic device to function as an over-driven amplifier in oscillator arrangements. A practical feature of logic circuits is that a number of them can be inexpensively utilized in oscillator configurations instead of just a single device.

A typical oscillator using logic circuits is shown in Fig. 5.17. Inasmuch as two cascaded NAND gates are used, the overall loop gain is very high and the square-wave oscillation is vigorous with steep rise and fall times. When

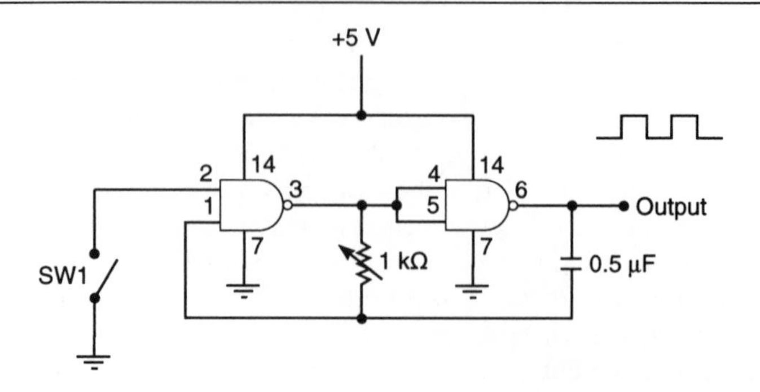

Fig. 5.17 *NAND element square-wave oscillator for the 1kHz region. Actually, there is only one pin 14 and only one pin 7 on the SN7400 module, which contains four NAND gates. Only two of these gates are used in the above oscillator*

we speak of 'loop gain', the reference is made to the feedback gain that exists within the very narrow 'linear' response of the gates. When the inputs of NAND gates are tied together, the devices function as inverters. This remains true even if one of the inputs is floating because it then assumes a high digital state. However, if one of the inputs is brought low, say by grounding, inverter action does not take place and the circuit is inhibited from oscillating. Thus, switch SW1 in Fig. 5.17 can be used to turn the oscillations on and off. This can be a useful feature in certain systems, especially when SW1 is replaced by an electronic switching technique.

Although an audio frequency application is depicted, operation may be extended to several megahertzs or higher by appropriate selection of the RC timing parameters.

A few words about the SN7400 NAND gate IC

The circuit, pin-connections and truth table shown in Fig. 5.18 provide a useful information for the use of the popular SN744 TTL quad NAND gate. Circuits configured about this, and other logic devices, generally omit power supply connections. However, the pin-connection diagrams shows that pin 14 is the + 5 V V_{CC} connection and that ground connects to pin 7. This technique simplifies schematics, enabling one to confine and focus interest on the essentials of circuits which often comprises numerous interconnections.

From the circuit of this device, one can infer that considerable gain is available, greater than can readily be obtained from a discrete device. Although such logic devices make speedy transitions between their on and off states, the gain during these transitions is available as feedback in oscil-

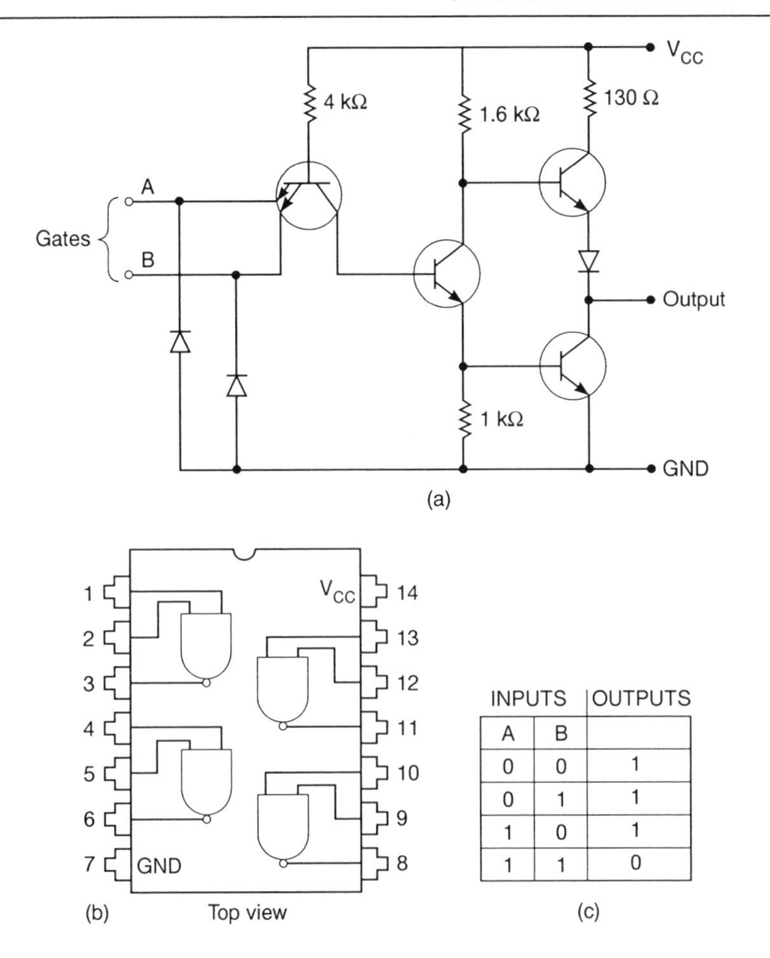

INPUTS		OUTPUTS
A	B	
0	0	1
0	1	1
1	0	1
1	1	0

Fig. 5.18 *Circuit, pin-out and truth table of the SN7400 NAND gate. This device is typical of logic ICs used in oscillators. a, Internal circuit of the SN7400. Note dual-emitter input transistor. b, Pin-out of the SN7400. Note common power supply provision. c, Truth table*

lator circuits. Even when crystals or LC resonant tanks are used for frequency determination and stabilization, the output of such logic devices as the SN7400 is a square wave. Fortunately, this is the required waveshape for clocking logic systems. For other purposes, the square wave may be converted to a sine wave with a resonant circuit or filter. One of the features of using logic gates for clock oscillators is that the voltage levels are compatible for the operation of the clocked logic systems.

The truth table reveals that the SN7400 becomes an inverter if the gates are tied together. Thus, the device will simulate the characteristics of a common-emitter transistor. An unconnected gate tends to assume the high state. Assuming, then, that such an unconnected gate is always high, the truth table indicates that the device will still work as an inverter. The oscillator of Fig. 5.17 should therefore operate in essentially the same manner whether the extra gates are unconnected or tied to the active gates. In actual practice it may pay to experiment with these connections especially at higher frequencies.

Logic circuit square-wave oscillator with crystal stabilization

The crystal oscillator shown in Fig. 4.19 embodies the same oscillation principle as the low frequency square-wave oscillator of Fig. 5.17. Thus, if the crystal is shorted out, a self-excited oscillatory mode occurs in which the frequency is governed by the 33 pF trimmer capacitor. In some instances, a higher free-running frequency is generated when the crystal is removed from its socket because of the stray capacitance of the holder. But this should not be inferred to signify instability when the crystal is in place; for then, the oscillation frequency is firmly that of the crystal. Inasmuch as there is no tuned tank circuit, other crystals than the 1-MHz unit indicated may be used. Regardless of the type of crystal used, series-mode resonance is involved in this circuit. Even an overtone crystal is likely to oscillate at its series-resonant fundamental mode. Feedback is very strong and crystals of marginal activity tend to be sure-fire in this simple oscillator.

The third NAND gate functions as an output buffer amplifier. It, too, operates in the off–on manner of logic devices and ensures a clean square wave for driving subsequent logic. At the same time, it increases the fan-out that would be available directly from the oscillator circuit itself.

An interesting and useful aspect of oscillators of this type is that they are rich in odd-harmonic energy. Thus, they are immediately useful as frequency markers. Also, by means of appropriate tuned circuits, a harmonic frequency can be 'sucked out' of the output waveform and amplified or processed for a specific application. This is slightly different technique than that provided by frequency multiplier stages in 'linear' radio frequency systems. This circuit should work well up to 10 or 15 MHz, which is close to the practical limit for non-overtone AT-cut crystals. Incidentally, even though the circuit generates square waves, the crystal oscillates sinusoidally.

A clock oscillator formed from cross-coupled ttl NAND gates

Cross-coupled gates are frequently encountered in logic circuits. This configuration is basically that of the family of the multivibrators used for

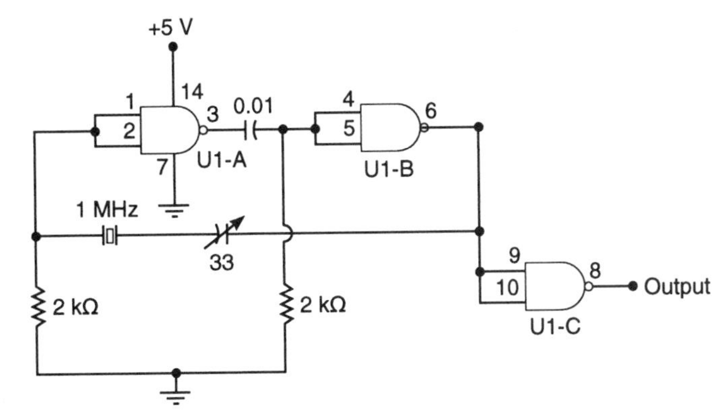

Fig. 5.19 *Crystal oscillator using NAND gates. The circuit includes a buffer output stage and delivers a square waveform*

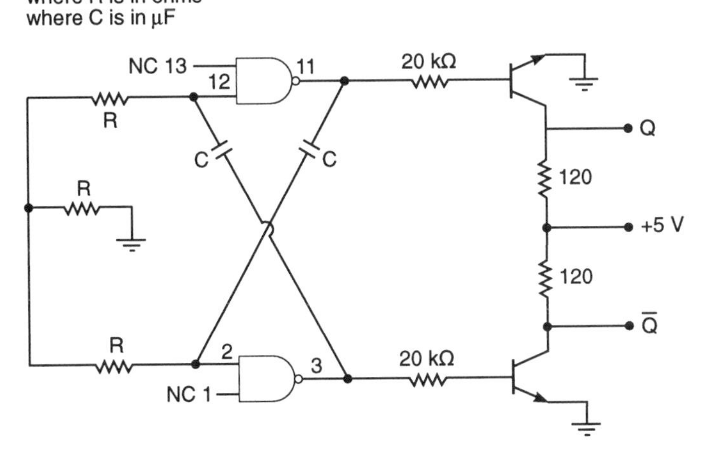

Fig. 5.20 *IC clock oscillator with complementary outputs. For proper operation, the R values should be in the vicinity of 100Ω. The C values have considerable latitude and should be selected to determine frequency*

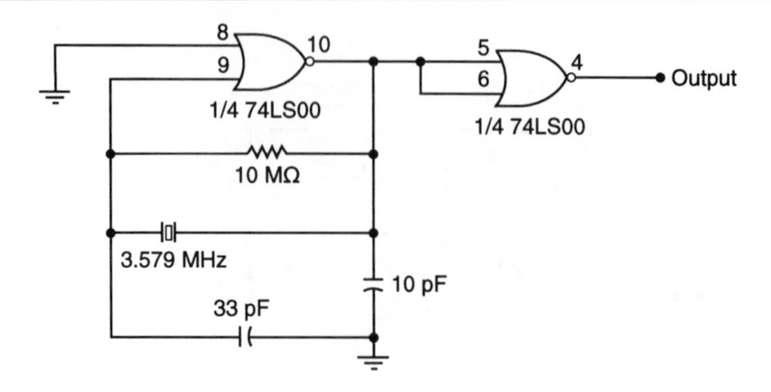

Fig. 5.21 *NOR gate crystal oscillator for supplying 3.579 MHz colour TV frequency. These crystals restore the sub-carrier frequency in TV sets*

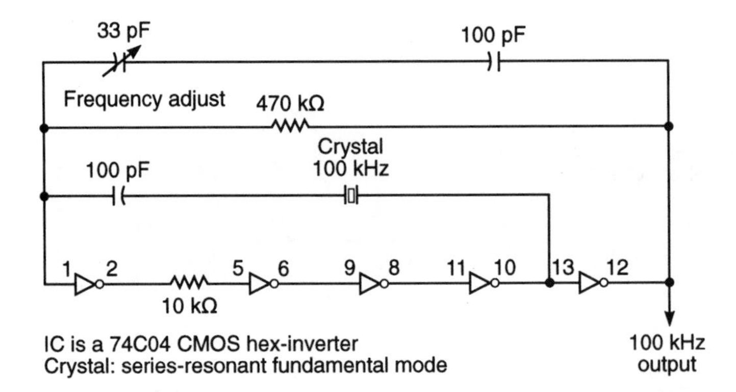

Fig. 5.22 *100 kHz oscillator using five sections of a CMOS hex inverter. Circuits of this kind employing several cascaded stages develop high feedback and can sometimes generate self-excited oscillations if the crystal is defective or is not in place*

many purposes with discrete devices. Logic gates often have very high gains and, if properly used, tend to be superior to discrete-device multivibrators. Thus, the arrangement shown in Fig. is not needed to square-up the signal, it provides buffer action and delivers a cleaner wavetrain than is available directly at the output of the first IC, i.e., the oscillator proper. It also provides increased fan-out capability by preventing needless loading of the oscillating portion of the circuit.

Another crystal oscillator arrangement is shown in Fig. 5.22. This one utilizes five devices in a hex-inverter IC. This is an exceptionally strong oscillator and is well-suited for frequency-marker generators. The frequency-adjust provision is somewhat unconventional: the trimmer capacitor, instead of being inserted directly in series with the crystal, operates in an

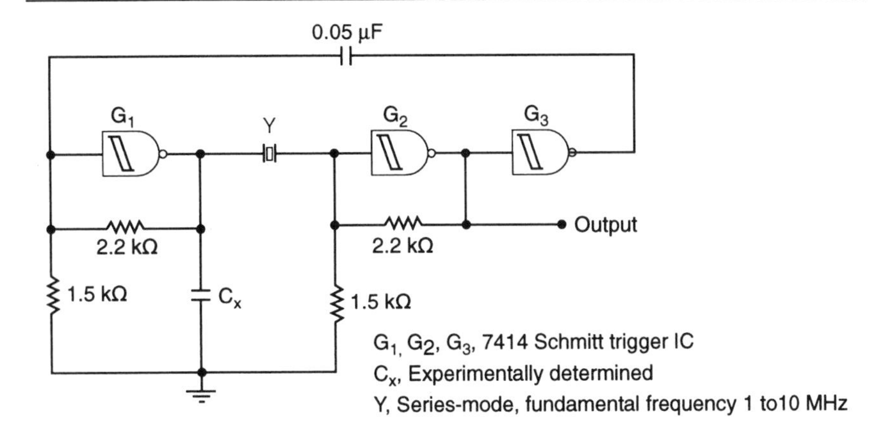

Fig. 5.23 *Crystal oscillator using Schmitt trigger logic elements. This is a vigorous square-wave oscillator. The function of C_x is to discourage generation of spurious frequencies*

overall negative-feedback loop. This tends to make the positive-feedback loop containing the crystal less sensitive to stray capacitance, thereby increasing the pulling range of the oscillating frequency. Spurious or non-crystal governed frequencies can be inhibited by experimentation with the resistor (10 kΩ) connected between the first and second stages. It should be noted that the last stage is not truly a buffer and it may be found profitable to utilize the sixth inverter in this IC as an output stage for better load-isolating action.

The arrangement shown in Fig. 5.23 is an interesting one employing Schmitt-trigger elements of a logic IC. Here, too, spurious frequencies may be excited by the vigorous feedback. A value of capacitor C_X may be empirically determined which will attenuate or inhibit spurious frequencies, but which will still allow reliable start-up of the intended crystal frequency. As with the previous oscillator circuit, it may be found desirable to add an additional stage to perform as a buffer amplifier.

The MOS NOR-gate oscillator depicted in Fig. 5.24 is especially useful for logic systems requiring complementary clocking. It will be seen that the output waveforms of this oscillator are displaced in phase by 180°. Insight into the nature of this oscillator scheme may be clarified by noting that the NOR gates behave as inverters because their gates are connected together.

Voltage-controlled oscillators

Voltage-controlled oscillators have many important uses. A few of these are found in phase-locked loops, instrumentation, such as the voltage-to-

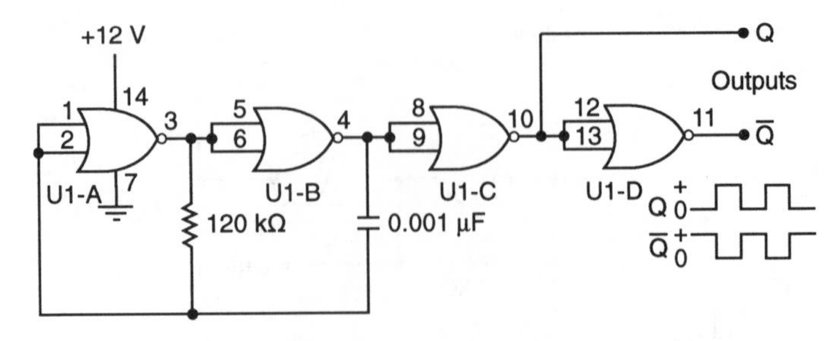

Fig. 5.24 *1 kHz complementary output oscillator using MOS NOR gates. The square-wave outputs are 180° out of phase*

frequency converters in some digital multimeters, tuning in communications equipment, certain switching regulators, telemetering and remote-control techniques. Two general approaches are common—one for radio frequencies and the other for audio frequencies. A voltage-dependent capacitor is used for the radio frequency range as part of a resonant circuit. Such a voltage-dependent capacitor is usually a reverse-based junction diode or varactor. Because negligible bias current flows, such a simulated capacitor exhibits low loss and the overall Q of the resonant circuit can be very high, being governed mostly by the inductor. This type of voltage-controlled oscillator has already been discussed. A typical circuit is that of Fig. 5.11.

For audio frequencies, varactor techniques can also be used, but there are certain practical difficulties, one being the outsized physical size of the tuned inductor. Another is the relatively low capacitance of pn junction diodes. Of course, there are low-frequency oscillators that do not use inductors, and it is possible to parallel a number of varactors. But most of the audio-frequency voltage-controlled oscillators encountered in practice make use of some type of relaxation oscillator or multivibrator in which either resistance or voltage level is manipulated by the control voltage. Figure 5.25 shows a unijunction oscillator in which the charging rate of the timing capacitor, C, is altered by a d.c. control voltage. Here, the bipolar transistor is connected to act as a voltage-dependent resistance shunting the timing capacitor. The control range extends from about + 0.8 to 2.5 V with the frequency decreasing with increasing control voltage. The component values are suitable for the mid-audio frequencies and the actual range of controlled frequencies is determined by selection of capacitor C.

An alternate way of voltage-tuning a unijunction oscillator is shown in Fig. 5.26. In this circuit, the bipolar transistor is in series with the timing capacitor and thereby takes the place of the 50 kΩ charging resistance used in Fig. 5.25. In this arrangement, the frequency increases with increasing d.c. control voltage. The biasing resistance, R_X can be empirically determined to produce an optimum relationship between applied voltage and resulting

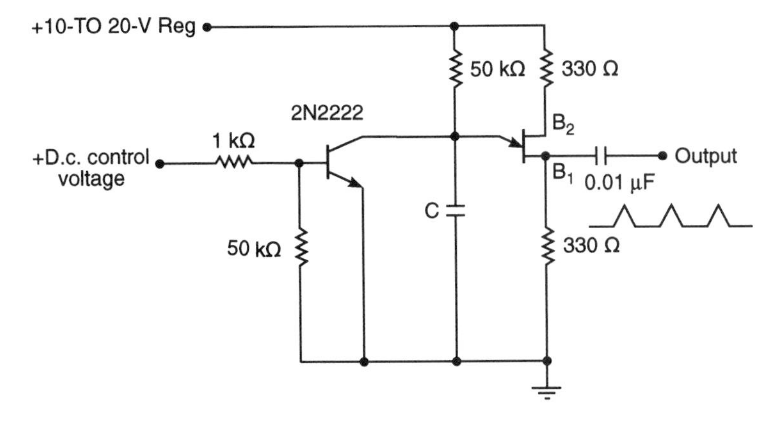

Fig. 5.25 *Simple voltage-controlled oscillator using a unijunction transistor. The bipolar transistor is used as a variable resistance to delay the charging rate of timing-capacitor C*

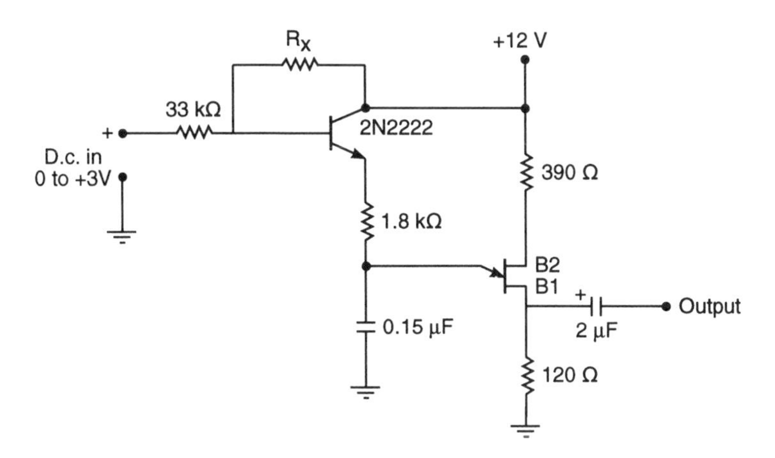

Fig. 5.26 *Another unijunction VCO circuit. In this arrangement, the bipolar transistor acts as a voltage-controlled resistance in the series-charging path of the timing capacitor. Experiment with R_x for best results; 330 kΩ is a good starting value*

frequency. Its value will depend on the beta of the 2N2222 transistor and the nature of the variable d.c. voltage source. The 2 μF output coupling capacitor enables the oscillator to drive a small 4- or 8-Ω dynamic speaker.

Another type of VCO (voltage-controlled oscillator) is shown in Fig. 5.27. This is essentially a multivibrator circuit much used for square-wave generation. Here, however, the basic circuit is modified to allow control of the voltage level that the single feedback capacitor 'sees' during its charge cycle. The higher the positive control voltage, the less time must be

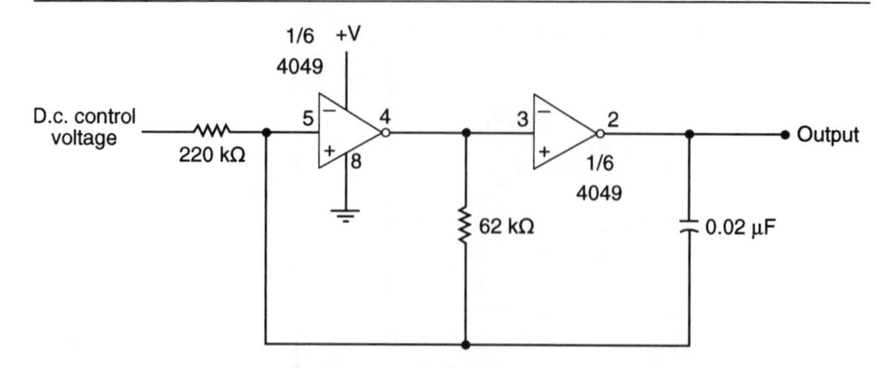

Fig. 5.27 *Audio frequency voltage-controlled oscillator using MOS inverters. This circuit techniques leaves four inverters for other circuit or systems functions*

consumed in the charging cycle and the higher is the output frequency of the oscillator. Some experimentation may be in order with regard to the two resistors in order to optimize the control range. Within limits, decreasing the value of the input resistor (shown as 220 kΩ) will improve the control range, particularly on the high-frequency side. It is always best to operate circuits of this type from a regulated d.c. source so that only the d.c. control voltage can change the frequency. Although MOS inverters are shown, the basic idea can be extended to other devices such as op amps and comparators. Also, it may prove feasible to use this scheme well into the moderate radio frequencies, say in the several-MHz range.

For applications where the output frequency of an oscillator must be a very linear and precise function of the d.c. control voltage, it is not wise to attempt to achieve this with discrete devices or even with op amps. Semiconductor firms now make dedicated ICs especially designed for such high performance. These may be called either voltage-controlled oscillators or voltage-to-frequency converters. Two such ICs that yield exceptional linearity are the National Semiconductor LM131 and the LM566. Also, Siliconix makes the LD111, a voltage-to-frequency converter intended for digital multimeters.

Another way to obtain better performance from a voltage-controlled oscillator than is readily feasible by 'rolling your own' is to make use of the VCO contained on the chip of a phase-locked loop IC. Fortunately, in certain ICs of this type, the VCO section of the loop may be used independently. A dedicated IC of this type is the 4046 CMOS phase-locked loop. Figure 5.28 shows how a simple VCO can be configured by appropriate selection of the IC pins. In this arrangement, the frequency range is governed by the RC time-constant of the resistor connected between pins 11 and 8, and by the capacitor connected between pins 6 and 7. With the indicated component values, the frequency is controlled from about 1 Hz to 10 kHz.

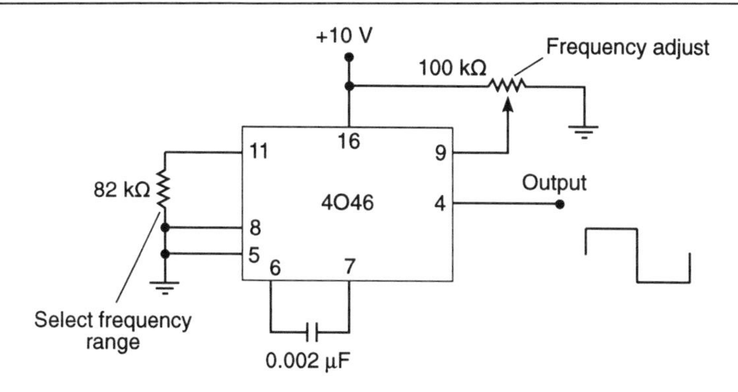

Fig. 5.28 *A VCO derived from a CMOS phase-locked loop IC. This dedicated IC provides pin connections so that the internal VCO may be used independently*

This may be extended to about 1 MHz by appropriate selection of the RC time constant. However, the resistive component of the RC value should be greater than 1-kΩ for reliable oscillation. Despite the tremendous control range provided by this sub-system, the integrity of the 50% duty cycle square wave is maintained.

The Schmitt-trigger oscillator

The relaxation oscillator shown in Fig. 5.29 has both practical and theoretical significance. Visual inspection reveals it to be about as simple a circuit as one could hope for. The oscillating mechanism is interesting inasmuch as it operates very much in the manner of a gaseous lamp oscillator. It will be recalled that the ionization-deionization process in the latter device gives rise to voltage hysteresis. Specifically, in the gaseous lamp oscillator, the ionization voltage is always higher than the deionization voltage. To prove that such a behaviour is necessary for relaxation-type oscillation, one has only to try to emulate such operation with a Zener diode. The Zener diode appears to be the solid-state analog of the gaseous lamp in certain respects: both devices hold off conduction below a certain voltage-level. Thus, both devices can be used similarly in voltage-regulator circuits and in circuits where voltage-clipping is the behaviour needed.

The Zener diode, however, exhibits zero voltage hysteresis. For all practical purposes, ionization and deionization voltages are identical and this disallows its use in RC relaxation oscillators. It turns out that the Schmitt-trigger device of Fig. 5.29 emulates the voltage-hysteresis of the gaseous lamp even though ionization phenomena are not involved in this simulation. Specifically, the Schmitt trigger delivers a high-impedance output as

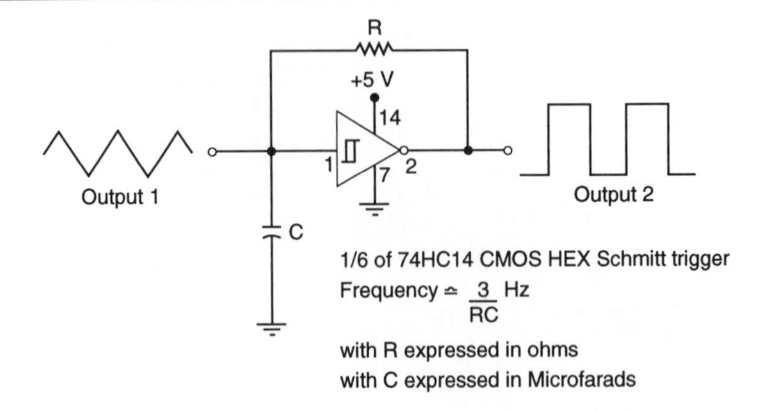

Fig. 5.29 *The Schmitt-trigger oscillator. The duty cycle is inherently 50%. Experiments can start with R = 100 kΩ and C = 0.01 μF, corresponding to a frequency of several kHz. The exact frequency will depend upon the upper and lower trip points of the Schmitt trigger. Triangular Output 1 is vulnerable to loading and is best sampled through a buffer stage*

long as the input voltage is below a certain triggering level. When this input voltage level is reached or exceeded, the output regeneratively snaps to a low-impedance level. Significantly, when the input voltage again falls, the device remains in its low-impedance state until the input voltage passes through a lower value, whereupon the Schmitt trigger abruptly returns to its 'off' or high-impedance condition. The voltage difference between the turn-on and the turn-off levels constitutes the voltage hysteresis of the device. The capacitor in Fig. 5.29 is thereby alternately charged and partially discharged, which is tantamount to saying an oscillatory process is taking place.

Note that two waveforms are available from this relaxation oscillator. The waveforms are more precise and stable than in a gaseous lamp relaxation oscillator. The regenerative snap actions of the Schmitt trigger are much faster than either ionization and deionization times in gases and are less vulnerable to environmental conditions. Whereas gaseous lamp oscillation is limited to audio frequencies, the Schmitt-trigger relaxation oscillator can be worked to at least several megahertz before waveform deterioration sets in. Artful control of stray parameters in the capacitor, the charging resistance and the monitoring technique could conceivably enable even higher-frequency oscillation via this simple scheme.

6 Special oscillator topics

In this chapter, it will be useful to reconsider some of the previously discussed oscillators in some greater detail; primarily, this will facilitate tailoring performance for specialized applications. For example, many applications in communications, instrumentation and scientific research benefit from oscillators that are both variable and very stable in frequency. As we know, these dual requirements represent an oxymoron. The use of a crystal oscillator is fine where applicable, and solution via synthesized oscillators is elegant, but costly (and often the source of noise problems). Accordingly, it will be found profitable to see how VFOs and VXOs can be optimized simultaneously for variability and stability.

Past techniques centred around electron tubes, silver-mica capacitors and temperature-controlled environments. The state of the art is improved now because solid-state devices develop relatively little heat. Moreover, tuning components are available with either low or precisely-specified temperature coefficients. Then, too, regulated d.c. power supplies have entered into common use. Last, but not least, low-cost frequency-counters now enable monitoring of frequency drift as it occurs; this makes it much easier to implement compensatory techniques. Obviously, the design and operation of such oscillators is both an art and a science.

We will see that there are many other specialized demands likely to be made on oscillators. These may pertain to frequency extremes, waveshape and purity, duty cycle, 'noisiness' and control. With regard to the last-mentioned performance parameter, we will find that turning an oscillator *on* and *off* is not always the simplistic control it sounds like.

We shall also find some unique associations between oscillators and heterodyning techniques that neatly lead to unusual but practical applications. Also, oscillators premised upon both analog and digital principles will be recognized to offer specialized applications to the creative designer. Finally, due attention will be directed to the often-ignored problem of suppressing *undesired* oscillations.

Guidelines for optimizing VFO performance

The construction of a very stable VFO generally requires some 'customizing' design changes once the circuit is put into operation. Because of complex interrelations between solid-state devices and various circuit components, optimum frequency stability is seldom obtained immediately after completion of the project. It will invariably be found that adjustments of excitation levels, coupling capacitors, biases and operating voltages will be called for. Often, different inductor types or configurations will pay off in tighter frequency constancy. Even bypass capacitors can sometimes be optimized for best results. When all these things have been done, one should not be surprised to find that slightly different results will be observed by experimenting with different solid-state elements of the same type.

To begin with, however, it is possible to estimate satisfactory performance by observing general guidelines. In so doing, it will be helpful to refer to the basic VFO circuit of Fig. 6.1.

1. Junction FETs and small-signal MOSFETS lead to good results more readily than bipolar transistors. The input capacitance of the bipolar transistor is relatively high and the resistive component of its input impedance is relatively low. Both components of its input impedance are quite temperature sensitive. This makes it difficult to obtain needed isolation from the resonant tank circuit. A bipolar transistor with a very high current gain or transconductance can, however, considerably alleviate this shortcoming. Gate connection techniques for FETs are shown in Fig. 6.2.
2. The time-honoured use of silver-mica capacitors in the frequency-determining parts of the circuitry is not always the best design technique. Actually, one finds an unpredictable range of temperature coefficients among otherwise identical capacitors of this type. What happens is that more experimentation is required than if one uses modern polystyrene or NPO ceramic capacitors. Polystyrene types have the interesting feature that their temperature coefficient just about cancels the temperature coefficient of certain ferromagnetic inductor cores. Additional insights into compensation via negative temperature-coefficient capacitors are provided by Figs 6.3 and 6.4.
3. At least one buffer stage should follow the oscillator for the sake of load isolation. A loosely-coupled source, gate or emitter-follower is best. Then, if further voltage or power level boosting is needed, this can be done in a second stage.
4. Plan on operation from a tightly-regulated, well-filtered power supply. Other things being equal, the lower the operating voltage of the oscillator, the better.
5. All bets are off if due consideration has not been given to the layout

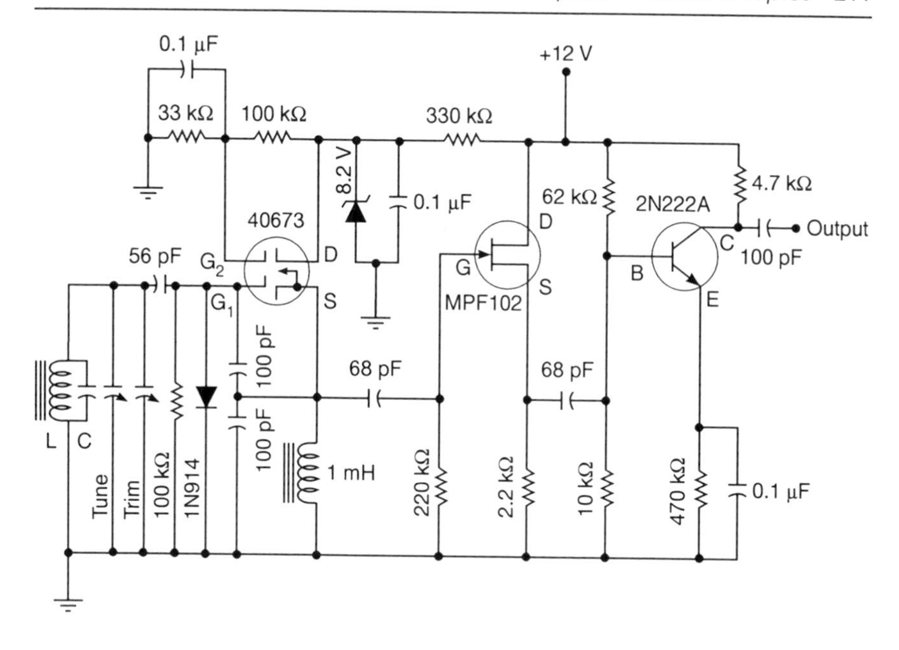

Fig. 6.1 *A basic VFO circuit. Select L and C for desired frequency-range. Good performance is readily attainable from 3 to 8 MHz. For many purposes, the TUNE capacitor can be a 100pF air-variable type. A 25pF ceramic type should suffice for the TRIM capacitor. Parameter values shown are for sure-fire operation. Modifications may be required for optimum frequency stability*

and the mechanical aspects of the oscillator. Sturdy hardware and good mounting techniques are mandatory to immunize the circuitry against vibration and physical deformation. Good shielding is needed, but there should be a respectable distance between the inductor of the tank-circuit and adjacent surfaces of the enclosure. The physical location of the oscillator should be where it is best isolated from any heat source. The mechanical and electrical integrity of tuning capacitors are exceedingly important. A high-quality tuning dial is required to ensure replication of a selected frequency.

6. After all of the above-listed matters have received adequate attention, one can refine frequency stability by experimenting with temperature-compensating capacitors in the resonant circuit. This entire procedure is analogous to the design philosophy of a really good audio amplifier. Cheap amplifiers simply incorporate a high amount of negative feedback which is supposed to overcome the consequences of an initially bad designer. Superior audio-amplifiers employ lower amounts of negative feedback because the initial design is quite good

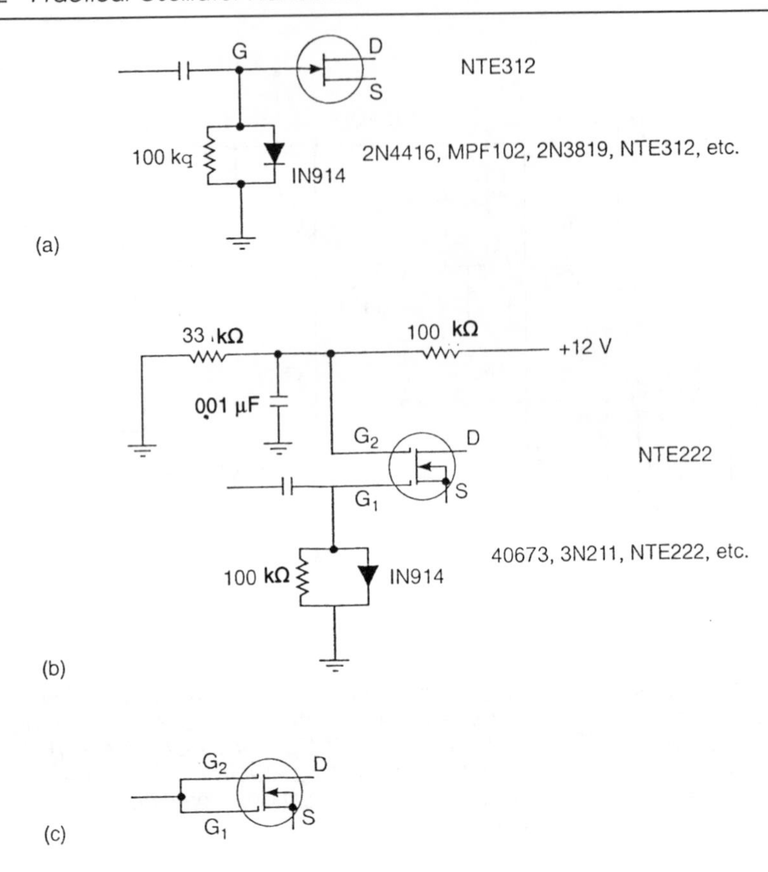

Fig. 6.2 *Methods of implementing field-effect devices in oscillators. a, JFET gate connections. The main purpose of the diode is to prevent forward conduction in gate of the JFET. b, Dual-gate MOSFET connection. The main purpose of the diode here is to limit oscillation amplitude in order to improve waveform and frequency stability. c, Alternative way to use the dual-gate MOSFET. This often amounts to a direct replacement of a JFET.*

already. The second approach, as audio enthusiasts well know, leads to flatter frequency response, lower distortions and more faithful reproduction.

7. It is easier to attain good frequency stability at lower frequencies than at higher ones. This often leads to the worthwhile practice of designing the VFO for a lower than needed frequency and then heterodyning to the higher needed frequency. Frequency multipliers can also be used for developing the high frequency. If the heterodyning technique is used, the oscillator would normally be a crystal because the overall

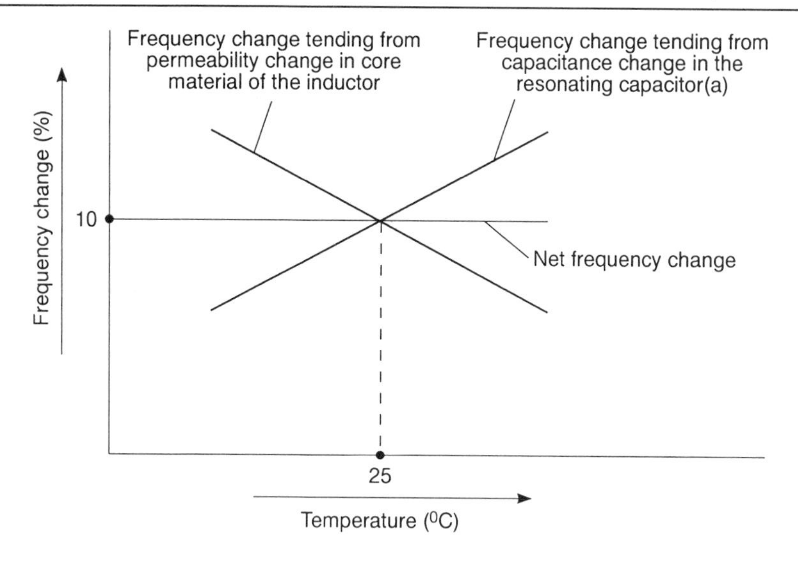

Fig. 6.3 *Idealized temperature compensation of frequency change. Core materials such as the much-used No. 6 and 7 iron-powder toroids exhibit a* positive *temperature coefficient. The frequency decrease with temperature that tends from this characteristic can be counteracted by appropriate choice of capacitors with a* negative *temperature coefficient. Precisely-characterized ceramic capacitors are often used, but sometimes polystyrene capacitors prove suitable*

frequency stability of such a system can be no better than its weakest link.

8. Those seeking the ultimate protection from frequency drift caused by temperature can provide a constant-temperature oven or heating arrangement. If so, the regulated temperature should be continually maintained, even when the VFO is not in use.
9. In theory, a mechanically-stable air inductor should be the best choice. Practically, problems arise because such inductors tend to require considerable space. This is not only because they have no high-permeability core, but because of their extensive electromagnetic field which couples to adjacent hardware and components. By the same token, such an inductor is very susceptible to fields from the power stages of the transmitter, the transmitting antenna and to other fields which may be present at the operating location. This can lead to frequency instability.

Iron powder toroidal inductors tend to be a good choice. In particular, the No. 6 material supplied by Micrometals Corporation and Amidon Associates exhibits exceptionally good temperature and ageing stability. These toroids are yellow-coded for easy identifica-

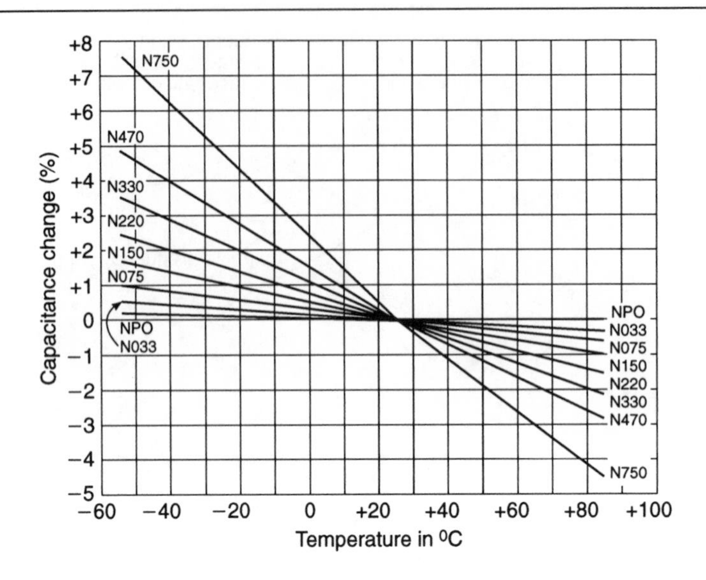

Fig. 6.4 *Standard selection of negative temperature-coefficient ceramic capacitors. Capacitors with appropriate capacitance and temperature coefficients can go a long way in reducing temperature-dependent frequency drift in oscillators. Note that the temperature coefficient of the NPO capacitor is zero. Experimentation often involves combining the NPO capacitor with one of the other. One strives to optimize frequency stability, and at the same time retain desired tank-circuit tuning*

tion. The permeability of mix 6 is about 8.5, and one can ordinarily expect good results over the frequency range of 2 to 30 MHz. The actual temperature coefficient is +30 parts per million per degree centigrade.

Another powdered-iron material, mix 7, is also worthy of consideration. Its permeability is given at 9.0 and its temperature coefficient is +30 parts per million per degree centigrade. These slightly better characteristics trade off, however, for the lower frequency range of 1 to 20 MHz. Palomar Engineers and Amadon Associates have been pioneers in the use of material No. 7, which is white-coded.

Some ferrite material have also been developed that look good for VFO applications. In particular, the Ceramag No. 11 material provided by the Stackpole Corporation exhibits a near-flat temperature coefficient over the useful range of 5 °C to 50 °C. Ferrite materials are more likely to show hysteresis effects if subjected to thermal abuse than powdered iron, but this should not be of consequences at the power levels encountered in VFO circuits. Go easy with the soldering iron, however.

Slug-tuned cores and cup cores have also been used with success. However, such configurations introduce the possibility of physical, and therefore, electrical instability. A good experimental technique is first to optimize a VFO operating with a toroidal inductor. Then, if the added tuning convenience of the slug-tuned inductor is desired, make the change. Any observed instability, drift or hysteresis effects can then be directly attributed to the slug-tuning hardware, and appropriate steps can be taken to reduce the effect. If continual tweaking of the frequency is not an anticipated feature, the remedy for such malperformance can be as simple as the application of Q-dope or epoxy to stabilize the core physically.

10. Other things being equal, strive for a high C-to-L ratio in parallel-resonant tank-circuits. One cannot go *too* far in this direction without restricting the available tuning range. In series-resonant tanks, such as in the Clapp oscillator, one strives for the opposite, a high L to C ratio. Here, one ultimately runs into the effects of irreducible stray and distributed capacitance. It is a good idea to consult published circuits of VFOs because others have also encountered the various limiting factors and have been forced to adopt practical component parameters. (If the limiting effects were not there, we could easily duplicate the performance of a quartz crystal by series resonating a very large inductor and a tiny capacitor!)
11. The tuning and trimmer capacitors are all too often unsuspected sources of frequency drift or instability. While many different conductive metals and dielectric materials are used in these variable capacitors, the common denominator is the need for mechanical adjustment. Thus, not only are these components subject to dimensional change with temperature, but there is the problem of reliable electrical contact with the moving element. Whereas a temperature coefficient can usually be largely cancelled out, hysteresis effects or erratic changes are not so easily remedied. The source of such malperformance is very often the variable capacitors.
12. A realistic procedure is called for in evaluating the frequency stability of the VFO. The first half hour or so after it is turned on, the frequency drift can be expected to be relatively large. This initial warm-up period is normal for even the best-designed oscillators. Thereafter, the drift should gradually slow down, the best attainable stability being realized in an hour or two. There is nothing sacred about any of the time periods cited—much depends upon the nature of the application. In some cases, the oscillator is best left on continually in order to avoid needless temperature cycling.
12. Previous methods of monitoring frequency drift were tedious and often inexact. Nowadays, radio receivers often provide digital read-outs of frequency. Also, frequency counters can be compact, low-cost

items. Because of these instrumentation techniques, the operation of the VFO can be instantaneously monitored and optimization of performance is greatly facilitated. Much of the guesswork in setting frequency has been dispensed with, and we are no longer at the mercy of parallex in reading dials or of adjustment hysteresis of the tuning elements.

14. After the above 'logical' cause-and-effect relationships have been investigated, do not be afraid to try for more refinement via other modifications. Sometimes ceramic bypass capacitors are best replaced by film types. Although quarter-watt resistors may be adequate from power dissipation calculations, half-watt types sometimes are found to improve temperature-dependent frequency drift. The excitation of Colpitts-family oscillator circuits often makes use of same-sized feedback capacitors. Improved stability may sometimes attend a modification of this one-to-one ratio. (Usually, the capacitor closest to ground is made larger in order to decrease excitation.) Be sure that the shield boxes containing the oscillator circuitry are sufficiently vented to retard the rise of temperature. Finally, couple out as little oscillator energy as is consistent with the required operation.

Some notes on VXOs

A few words regarding variable-frequency, crystal oscillators (VXOs also known as VCXOs) are appropriate following our guidelines for optimizing VFOs. This is because 'pulling' the natural frequency of the crystal resonator gradually degrades the mode of oscillation so that the frequency is more and more self-excited in nature, and less and less crystal-stabilized. When you read in the literature that a certain VXO circuit can provide such-and-such tunable frequency-range, it is not clear just how much stability is contributed by the crystal at the outer limits of this range. By this same reasoning, one expects the highest stability when the crystal alone governs the oscillation frequency, i.e., when the frequency isn't being 'pulled'.

In the light of the above, it stands to reason that regardless of other aspects of the VXO circuit, it should be capable of functioning as a very good VFO. If such design and implementation is kept in mind, a worthwhile extension of the adjustable frequency range will invariably be realized. Accordingly, one should abide by the guidelines listed for VFOs. Thus, in the broadboard experimental stage, the VXO should first be operated and optimized as a temperature-stabilized VFO. For practical purposes, this can be done at a frequency somewhat removed from the 'reasonably' anticipated pulling limit.

In earlier discussions in this book, it has been already pointed out that VXO performance is very much an art and quite dependent on experimen-

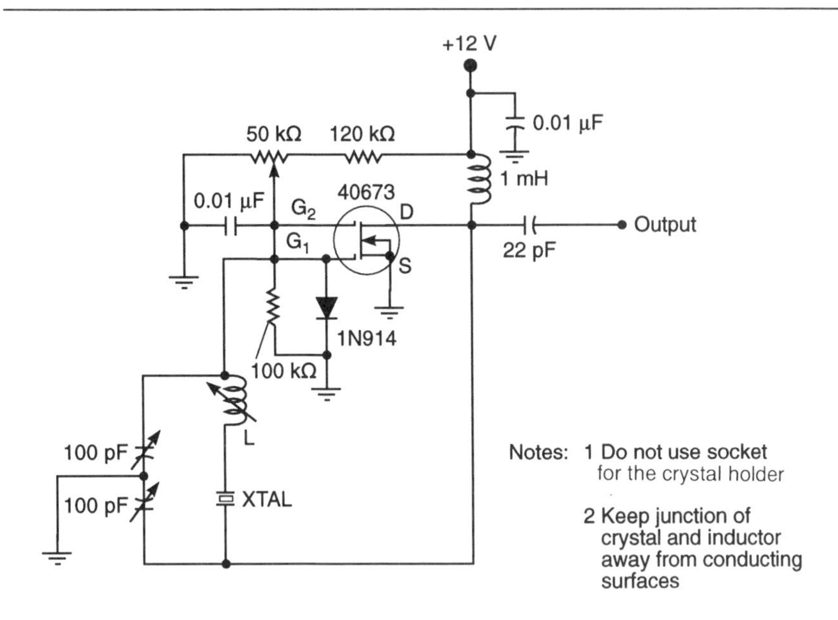

Fig. 6.5 *A basic VXO circuit. Derived from a Pierce oscillator, this VXO is capable of exceptionally wide frequency 'pulling'. It should be used with buffer amplifiers such as previously shown with the VFO. The slug-tuned inductor, L, should be selected for either a 7 MHz or a 3.5 MHz crystal. Adjust all variable elements for optimum pulling of the crystal frequency*

tal techniques. Other things being equal, AT-cut crystals in HC-6 holders are often found to provide the greatest frequency 'stretch'. It is necessary to reduce stray capacitances to a minimum and a socket should not be used with the older. Higher-frequency crystals tend to provide more than proportionate frequency tuning-ranges than do their lower-frequency counterparts. The frequency pulling effect is such that most of the tuning range is from the 'rated' frequency to a lower limit. The circuit of Fig. 6.5 might provide about 20 kHz of frequency 'pull' for a 7 MHz crystal. Greater range may be realized sometimes, depending upon the crystal's activity, the transconductance of the active device and the inherent quality of the circuit as a VFO.

Although only the VXO proper is shown in Fig. 0.5, buffer amplifiers such as previously depicted for the VFO should be used in practice. This is because the VXO is very vulnerable to frequency pulling from load changes. Interestingly, the frequency range goes up proportionately with frequency multiplication. Thus, by following a low-frequency VXO with suitable frequency multipliers, a range of, say, 150 kHz might be attainable at the amateur 2 metre band.

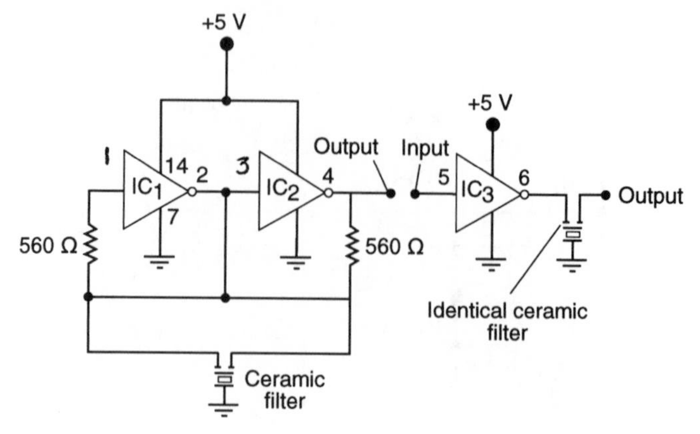

Fig. 6.6 *A ceramic-filter oscillator circuit. Shown also is an optional stage for further purifying the output waveform. (Input bias-adjustment can make this stage a tuned receiver.) The salient features of this 'flea-power' oscillator are its simplicity, low cost and compactness. The two commonly-available frequencies are 455 kHz and 10.7 MHz*

The ceramic filter oscillator

Ceramic filters have been a successful as substitutes for double-winding IF transformers in superheterodyne receivers. These devices operate on the piezoelectric principle where the electrically induced vibrations from an input element acoustically couple to an output element, thereby developing an electrical signal. Frequency selectivity occurs in an analogous way to the same phenomenon in the more familiar quartz crystal. Although ceramic filters lack the precision of quartz crystals or resonators, they nonetheless perform as well or better than the tuned transformers they replace. Moreover, they are very compact, as well as being cost effective. Unlike quartz crystals, ceramic filters and resonators are not available in a wide range of centre frequencies. Rather, the two commonly-available frequencies are 455 kHz and 10.7 MHz, since these correspond to the much-used IF channels of TVs and radio receivers.

It is only natural to ponder whether these piezoelectric devices are usable in oscillators. It turns out that they are very active and are readily excited at their natural oscillation frequency. Because of their less than compelling specifications regarding available frequencies, temperature coefficient and operating Q, a second question arises concerning applications for such an oscillator. Aside from the novelty aspect, it seems reasonable to suggest such uses as toys, test instruments, security systems, 'air-coupling' techniques for achieving electrical isolation and short-distance telemetering.

The circuit for a ceramic-filter oscillator is shown in Fig. 6.6. An additional stage is also shown and is useful if improved waveshape is desired. In any event such oscillators are 'sure-fire', with no need to experiment with L or C values. If required, however, the oscillation frequency can be 'pulled' by changing the resistances. It is interesting to note that if the ceramic filter is replaced by a small capacitor, a strong relaxation-type oscillation is produced. This mode of oscillation can sometimes be produced by a defective or ungrounded filter because of the capacitance between the input and output. Such oscillation will generally be far from the ceramic filter frequency.

Obviously, this circuit belongs to the same oscillator family as the several quartz-crystal oscillators employing various logic-gates which were discussed in the previous chapter. It is quite likely that, because of the expansion of telemetering, paging and intercom applications, that we may soon see ceramic resonators at other than the two traditional IF channel frequencies. Because of the inherent 'flea-power' of these oscillators, FCC licensing is not likely to be involved.

The regenerative modulator—is it an oscillator?

The regenerative modulator is a little-known circuit that was around during the earlier phases of vacuum-tube technology. In particular, it was used in carrier frequency telephony to achieve frequency division. When revamped with modern solid-state devices, this circuit technique offers interesting possibilities to experimenters. Basically, it provides jitter-free division of a sinusoidal input frequency, delivering a sinusoidal output which conveniently disappears if the input is interrupted.

So far, we do not seem to be describing an oscillator and, admittedly, there is much room to debate that it is not an oscillator. On the other hand, as will be shown, it operates much like an oscillator in some respects. In particular, feedback of energy sustains its operation.

The block diagram of the regenerative modulator is shown in Fig. 6.7. Note that a feedback signal heterodynes with the input frequency to produce the output frequency which, in turn, is appropriately multiplied to keep the process going. Let n represent the frequency dividing factor; that is,

$$n = f_{in}/f_{out}$$

Then, in order for the circuit to operate the feedback frequency must be $(n - 1)\, f_{out}$. If this requirement is met, the resultant difference frequency emerging from the mixer will reinforce the process and keep it going. It is only a matter of properly selecting resonating frequencies of the tank circuits. If you experiment with integral values of n, it will be seen that it is easy enough to calculate the needed resonant frequencies. n values up to

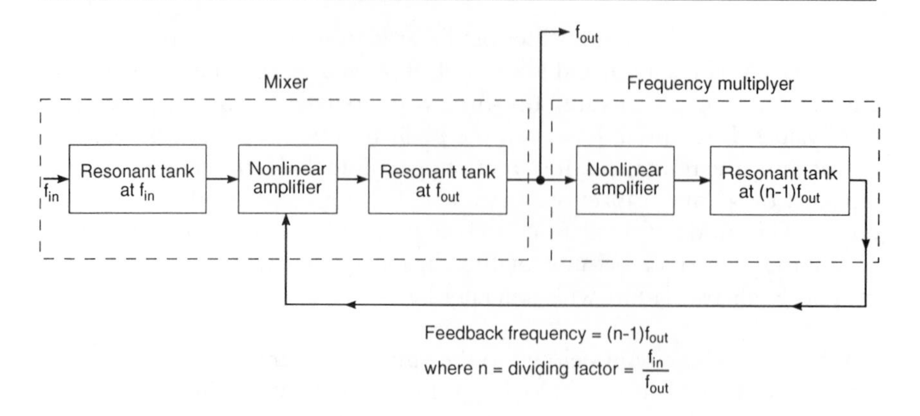

Fig. 6.7 *Block diagram of the regenerative modulator. Although an internal frequency-multiplication is performed, the overall operation is frequency* division. *Like a true oscillator, such an arrangement requires energy feedback for its operation*

about ten are not difficult to work. Beyond ten, special attention must be given to achieving high operating Qs to ensure reliable output of the desired sub-harmonic and rejection of the adjacent subharmonics. A basic circuit is shown in Fig. 6.8.

The overall efficiency of this circuit depends upon how much nonlinearity or overdrive can be produced in the two stages, this being a requisite of both mixing and frequency multiplying. By employing crystal or ceramic resonators in place of LC circuits, the experimenter can attain some inordinately-high division factors. Possible applications include operation on the amateur 'top' band (160 metres) from already extant 80 metre equipment, and the production of intermediate frequencies from reception of satellite microwave signals. One could also conjecture a phase-locked loop (PLL) in which a regenerative modulator could replace the conventionally-used frequency counters. Such a PLL could conceivably be electrically quieter because all internal signals would then be sinusoidal.

The phase-locked loop and synthesized oscillators

The phase-locked loop, as with the regenerative modulator, may be argued to be something other than a true oscillator. Actually, it is a system, rather than a single functional block. Moreover, technical literature does call it a synthesized oscillator, at least for some applications. And when we contemplate its operation, we are reminded of a VFO, one usually endowed with much greater precision and flexibility than either VFOs or VXOs. Also, adding even more credibility to our contention it does not unduly stretch

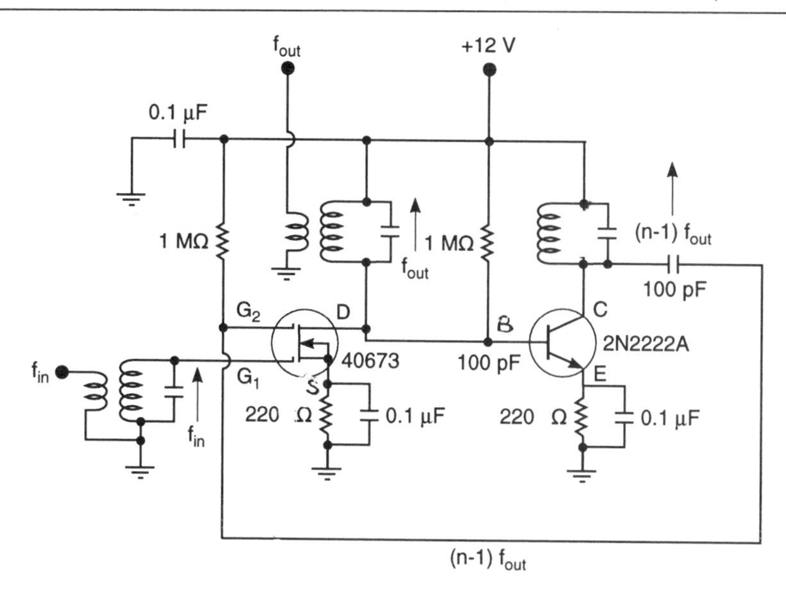

Fig. 6.8 *Basic circuit of a regenerative modulator. A wide spectrum of audio and radio frequencies can be readily accommodated. LC resonances and parameter values are not critical. If sufficient drive-level exists, the operation is reliable and jitter-free. Unlike flip-flop frequency dividers, this circuit delivers a sinusoidal output*

the imagination to see the phase-locked loop as a 'beefed-up' VCO inasmuch as one of its important building blocks is actually a VFO.

There are many applications of the PLL principle. In communications it readily provides multiple precise frequencies from a single, fixed-frequency reference frequency. PLL techniques actually substitute a single circuitry block for multiple crystal oscillators. PLLs are also used as FM and AM demodulators. In industrial control applications they function as servo-amplifiers and provide a neat way to control motor speed. PLLs can be made to incorporate analog or digital techniques and most make use of both. They are commonly found in applications covering the frequency spectrum from 60 Hz to the microwave region. At one time, the implementation of PLL circuitry was frought with technical problems and with high cost. However, the advent of numerous digital and analog integrated circuits makes most applications relatively straightforward, cost-effective and reliable.

The topic of PLL is a voluminous one and no attempt will be made to do other than bestow an explanation of the basic concepts. In its simplest form, the block diagram of the PLL is shown in Fig. 6.9. Briefly, this is a closed loop nulling system in which an error signal develops to make the phase detector always 'see' the same frequency at its two inputs. In the simple arrangement of Fig. 6.9, this implies that a d.c. correction voltage emerges

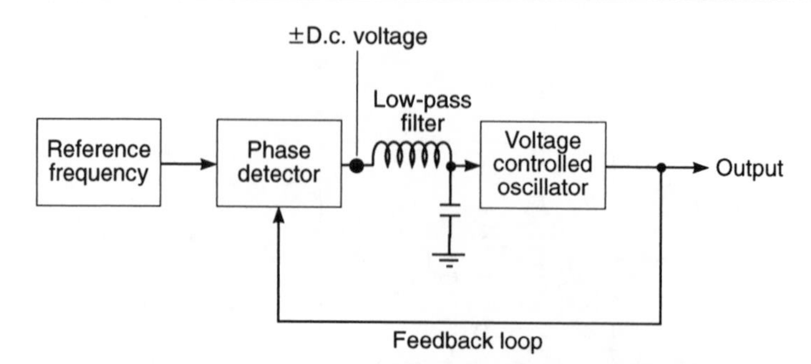

Fig. 6.9 *Block diagram of simple PLL. Ideally, the output of the phase detector will be 0 V and the output frequency of the voltage-controlled oscillator will be identical to that of the reference frequency. Then any tendency for the voltage-controlled oscillator to change its frequency will be counteracted by the production of a d.c. error voltage from the phase detector tending to restore the output frequency. Practically, a constant hunting of the nulling condition will attend operation*

from the phase detector which then returns the VCO frequency to where it should be to achieve nulling at the phase detector. In this set-up, this means that the VCO frequency will be automatically stabilized at the reference frequency. The corrective action is analogous to that occurring in a simple series-regulated d.c. power supply, except that a.c. rather than d.c. signals are involved in the PLL.

A very important feature of the PLL is that a wide range of precise frequencies can be provided by the system as long as the phase detector can 'see' identical frequencies at its two inputs. The significance of this statement can be gleaned from inspection of the slightly more complex arrangement shown in the block diagram of Fig. 6.10.

Let us explore some of the practical ramifications of the PLL system of Fig. 6.10. First, we see a frequency divider interposed between the phase detector and the reference oscillator. It happens that high-stability reference oscillators are relatively easy to design and construct for crystals in the 3 MHz to 8 MHz range. On the other hand, phase detectors often operate in the 1 kHz to 10 kHz range. This is dictated both by the nature of the phase detector and by the incremental spacing of communications frequencies. Therefore, a counter or divider is used to reduce the reference frequency to the lower frequency need at the phase detector.

On the other hand, the voltage-controlled oscillator (VCO) may operate from several MHz to several hundred MHz. At the higher frequencies, it is often burdensome from both cost and technological standpoints to obtain good performance from the programable frequency divider. So, here too, a fixed frequency divider or prescalar is employed to reduce the VCO frequency. Keep in mind that, despite the many circuit sections of the PLL

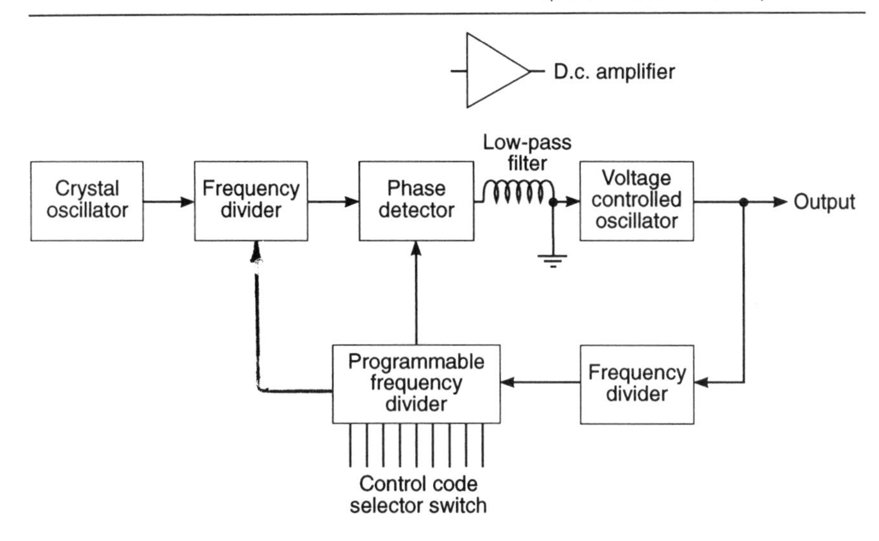

Fig. 6.10 *Block diagram of a practical PLL. A set-up such as this allows manual selection of many stabilized frequencies. Also, the individual functional-blocks operate within constraints favourable to optimum performance and cost considerations. The symbolically-depicted low-pass filter is often an RC type. A d.c. amplifier may follow the phase-detector*

system, as long as the phase detector can 'see' two nearly equal frequencies at its inputs, the system is capable of attaining lock and thereby providing a multitude of different frequencies, all stabilized to the reference frequency. Note also, that from the phase detector's 'viewpoint', the reference frequency is the divided down crystal-oscillator frequency.

There are many interlinked performance parameters of such a system. Other things being equal, we would want to endow the PLL with the ability to lock from a considerable departure from the reference frequency as seen by the phase detector. And, once locked, a forceful departure from the reference frequency should not readily break the tendency for restoration or mulling. Moreover, the settling time of the restorative operations should be short. Contradictions arise when there is an attempt to optimize all operating parameters. Experimentation with the low-pass filter and with operating levels is often needed to achieve the best balance of performance features. Often, one must also be concerned with wave purity, phase noise and adjacent channel interference.

The apparent complexity of the PLL-based frequency synthesizer is deceptive because all the functional blocks are readily available as low-cost integrated circuits. In some instances, a single IC suffices. On the other hand, it may sometimes be desirable to assemble one or more functional block from discrete circuitry. For example, VCOs of the type shown in Fig.

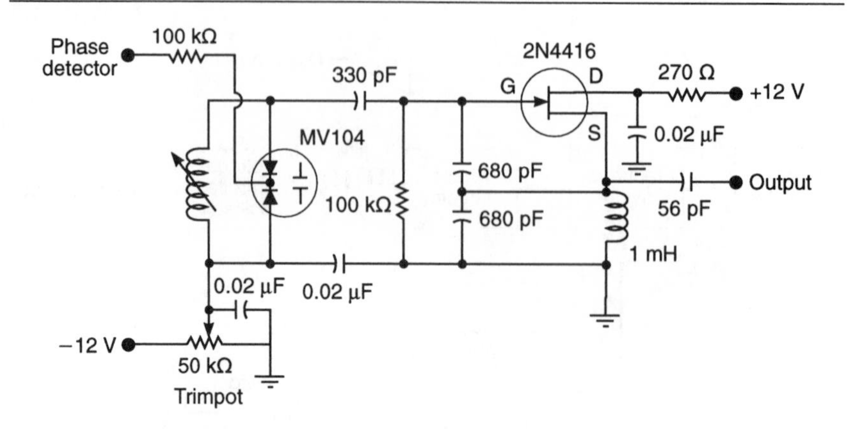

Fig. 6.11 *Voltage-controlled oscillator suitable for use in a phase-locked loop. Note the double varactor in the tuned circuit. Select slug-tuned inductor for the 3 MHz to 8 MHz region. (Much higher frequencies can be designed for with a bit of experimentation and care.) The trimpot provides the mid-operating bias for the varactors*

6.11 are often used. Note that this circuit differs from that of Fig. 5.11 in that two back-to-back varacters are used. This yields better performance because neither varacter can go into forward conduction.

Once the basic operating principle of the phase-locked loop is understood, one of its salient features will be seen to be its flexibility. As long as the phase detector has the opportunity to see equal frequencies at its input terminals, many circuit variations can be accommodated in the several building blocks to the system. Thus, certain applications will make use of frequency multipliers as well as dividers. Additionally, desired lower frequencies can be achieved by heterodyning against a stable oscillator. (Another analog method of lowering frequency would be through the use of the previously-described regenerative modulator.) Also, not to be overlooked is the use of a d.c. amplifier between the phase detector and the VCO.

There are many analog- and digital-type phase detectors. Keep in mind that the phase detector need not deliver zero volts d.c. when the system is nulled so that equal frequencies are represented at the inputs of the phase detector. It is only necessary that a suitable range of d.c. voltages be available for an appropriate VFO. For example, so-called 'zero error-signal' might actually be several or more volts at either polarity. Then, any deviation from this value would be properly identified as the real error signal. In some PLL applications, particularly those found in receivers, a single IC, such as the NE561 or NE565 is used and there are no LC circuits. Low pass filters in such instances are RC networks and VCOs are voltage-controlled multivibrators. Also, the phase detector in these ICs is a balanced mixer comprised essentially of active devices and having no need of tuned circuits.

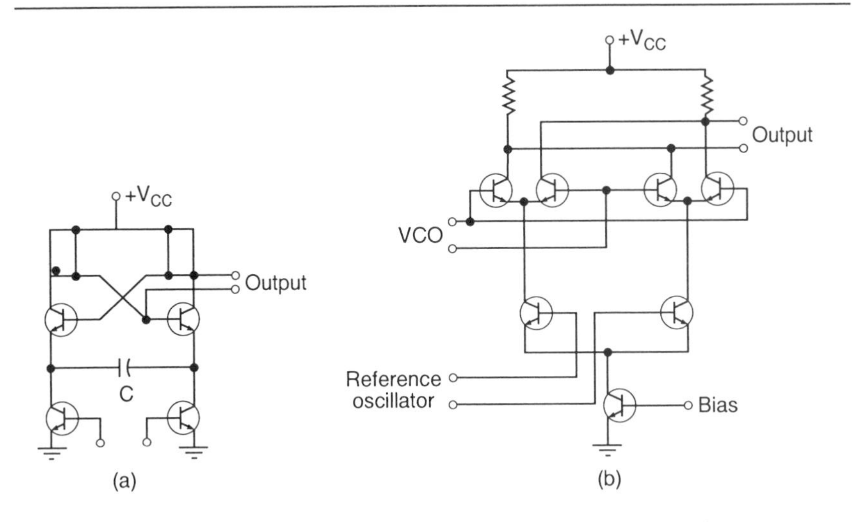

Fig. 6.12 *Simplified circuits of two of the function blocks within the NE560 PLL IC. Integrated circuits greatly simplify phase-locked loop systems. a, The voltage-controlled oscillator. The nominal frequency of this multivibrator is set by external capacitor, C. The control voltage varies frequency by changing the emitter currents of the multivibrator. No tuned LC circuits are needed. b, The phase detector. This is a double-balanced mixer making use of differential amplifier pairs. The d.c. component of the output is fed through a low-pass filter (and sometimes a d.c. amplifier) to the VCO*

The basic active circuits used in the NE561 and NE565 ICs are shown in Fig. 6.12. Other 'bells and whistles' are included in order to endow these ICs with applications versatility. For example, PLLs are used as FM demodulators, signal-tracking filters and synchronous AM detectors. With regard to FM demodulation, it is interesting to note that the recovered error signal represents the FM modulation on the RF carrier.

Although the experimenter and designer should have some insight into the nature of the functional circuit-blocks comprising a PLL system, it obviously makes good sense to go the IC route insofar as possible. For otherwise, the dozens of active devices, diodes and associated components can all too easily present problems of cost, reproducibility and reliability.

A second way of synthesizing frequencies from a reference oscillator

Another type of synthesizer also provides numerous stabilized frequencies from a single reference oscillator. For lack of a better name, let's call the second scheme a frequency synthesizer. The frequency synthesizer is el-

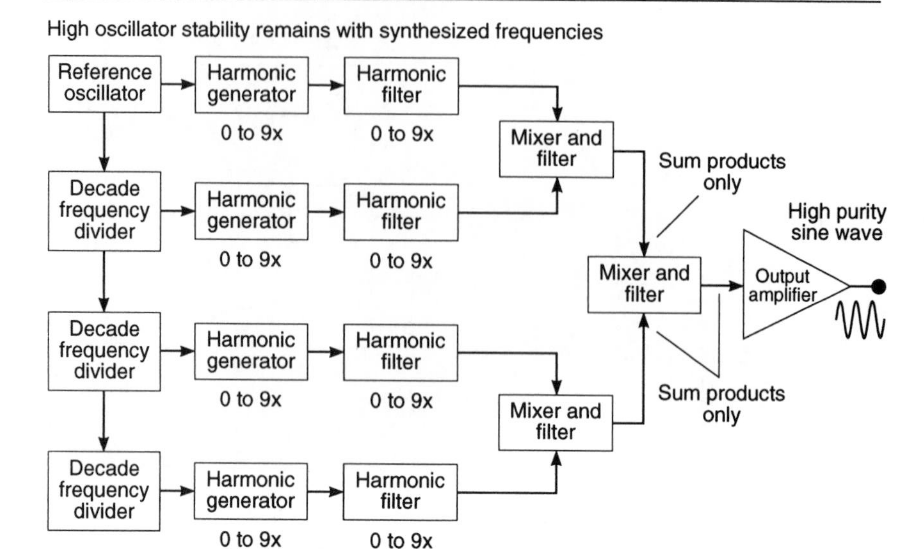

Fig. 6.13 *Simplified block diagram of frequency synthesizer using additive techniques. Frequency selection from panel switches is quite complex and is not shown. The decade frequency-dividers are formed from cascaded (and permutated) flip-flops. The harmonic generators generally use step-recovery diodes to produce a 'comb' spectrum rich in harmonics. The mixers produce sum and difference frequencies, but the filtering techniques reject the difference components. In this way, the output eventually becomes the summation of the various panel-selected decimal digits*

egant, but its complexity generally excludes it from consideration by experimenters and most designers. Rather, it is usually acquired from a specialized instrumentation company. Despite this situation, the basic principle of this oscillator system is rather simple. All that happens is that a number of harmonics and subharmonics of the reference frequency are produced, then selectively combined in an additive fashion to yield a desired output-frequency. This is easily grasped, but the demands made on the sorting out of the great number of combinations, and on filters to attenuate unwanted frequencies, are exceedingly great. Remarkedly, one can select from panel push-buttons, or thumb-wheel switches, numerous closely-spaced frequencies comprised of seven, eight or more digits. This is as easily done as punching out any desired number on a calculator.

Although specific circuitry details will not be covered here, the block diagram of Fig. 6.13 provides some insight into this oscillator system. Keep in mind that the bringing together of two or more frequencies in a mixer always generates *both* sum and difference frequencies. It is the *sum* that is desired and the difference frequency must be greatly attenuated. It is this filtering problem that mainly endows the scheme with its design complexity. At the same time, it should be obvious enough that once we have the liberty of additively combining a large selection of both subharmonics and harmonics, numerous output frequencies become available. Since the constituent frequencies all derive from the reference oscillator, they are inherently phase-locked to one another. This fact facilitates the combining process. Thus, if the reference frequency is 1 MHz and we wish to provide an output frequency of 100.45 kHz, one of the constituent frequencies would be the tenth-subharmonic of 1 MHz, or 100 kHz. This would be produced by a decade frequency-divider. The next constituent frequency would be the fourth-harmonic of the thousandth-subharmonic of 100 kHz. This requires three decade-dividers to reduce 100 kHz to 100 Hz. From the 100 Hz frequency, we can then obtain the requisite 400 Hz from a frequency-multiplier circuit. If the 400 Hz is next added to the 100 kHz in a mixer, one of the frequencies (the sum) produced is 100.4 kHz.

To get to 100.45 kHz, the above-outlined process is continued with appropriate decade frequency dividers and frequency multipliers, then combining again in a mixer to produce the sum frequency. Admittedly, a bit of tedious arithmetic is involved, but the basic idea is simple enough. Practical implementation is the headache.

Quelling undesired oscillations

Another matter closely associated with oscillators is the occurrance of unintended or undesired oscillation. In general, amplifiers have the potential of inadvertently becoming oscillators. This is because there is always some unintentional feedback path, as well as stray inductance and capacitance in practical circuits. Tuned amplifiers can often be easily pacified insofar as tendency toward oscillation at the resonant frequency of the main tuned tanks circuits. Oscillation at frequencies far removed from the main resonances are not so easily stopped and often requires a good deal of empirical effort. Such oscillatory behaviour goes under the general name of parasitics and is usually at very much higher frequencies than the intentional resonances. However, certain types of parasitic oscillations can also exist at lower frequencies than would be due to the intentional tank-resonance. Parasitic oscillations are undesirable because they extract additional power from the active device. Also, they convert this power into an energy format which often destroys components associated with the amplifier.

Power amplifiers for radio frequency transmitters using vacuum tubes have long been examples of circuitry where one of the criteria for successful operation is the elimination of unintentional oscillation. Unfortunately, it is entirely possible for such an amplifier to yield passingly-good performance at its designed frequency, but simultaneously also generating parasitic oscillations. Usually, however, such operation is not destined to go on indefinitely. Sooner or later, one finds that tube life is mysteriously shortened, or that radio frequency chokes, meters or other components are strangely destroyed, or that RFI at far-removed frequencies is traced to the amplifier.

To begin with, tubes, especially triodes, have inherent capacitance between their elements. The plate-grid capacitance is especially troublesome because it provides a positive feedback path directly from the output back to the input of amplifier circuits. Such a feedback path, it will be recalled, is one of the requisites of an oscillator. Because of the electrical characteristics of tuned circuits, another requisite is all too easily met—this is the phase condition wherein an appreciable part of the fed-back signal is in phase with the input signal. The input signal here may arise from thermal-noise fluctuations which though initially feeble quickly build up in magnitude from the reinforcing effect of the positive feedback. In this way, our amplifier becomes an oscillator. The unintentional feedback can support both the tuned frequency and parasitic frequencies from stray reactances.

The frequently used neutralizing techniques are exemplified in the circuits shown in Fig. 6.14. The two circuits for single-ended amplifiers work in very much the same way. In both instances, the neutralizing capacitor is adjusted to approximately the plate-grid capacitance of the tube. With that done, there is then cancellation of the positive-feedback energy that would otherwise occur through the plate-grid capacitance. It is interesting to observe that these amplifier circuits are basically tuned-plate/tuned-grid oscillators without the neutralizing provisions. Although neutralizing is primarily beneficial is quelling the tendency for oscillation at the resonant frequency of the amplifier, it does not necessarily inhibit parasitic oscillation at very high frequencies which result from the complex effects of lead lengths and stray parameters. To knock out the parasitics, another technique is commonly used. (Incidentally, frequency multiplying amplifiers rarely need neutralization because their grid and plate circuits are not tuned to the same frequency. For somewhat subtle reasons, however, some have claimed that these frequency multipliers operate more efficiently if they, too, are endowed with a neutralization network.)

Note the impedances, 'Z', in the circuits of Fig. 6.14. These are used in order to suppress the tendency for parasitic oscillations to develop. The idea is to have Z appear as a high impedance with high losses to the parasitic frequencies, but to represent relatively low impedance and energy loss at the tuned frequency of the amplifier. Commonly, the resistive component of Z will be a non-inductive resistor in the 10–100 Ω range. The inductive component of Z will then be formed from the effect of several turns of wire around the resistor. As depicted, these resistive and inductive components of Z are in parallel. It is easy to see that at a low or moderate frequency, the resistance will be essentially shorted out by a small amount of inductive reactance. Conversely, at the higher frequencies of parasitic oscillations, the

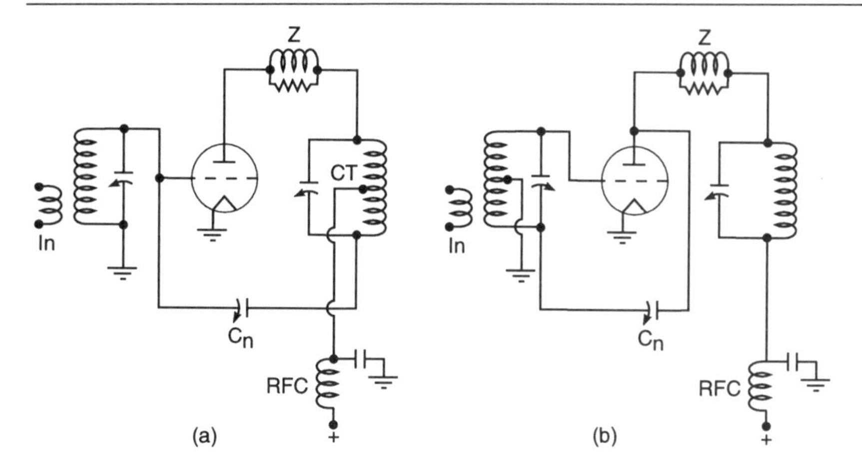

Fig. 6.14 *Simplified neutralizing circuits in electron-tube radio frequency amplifiers. Both techniques apply phase-reversed voltage to the internal feedback-path, thereby cancelling the positive-feedback from the plate-grid capacitance. C_n is the neutralizing capacitor, set to approximate plate-to-grid capacitance. a, Plate neutralization circuit. b, Grid neutralization circuit*

inductive reactance will be very high and this radio frequency energy will then be forced to dissipate itself in the resistance. One may say that this grossly overloads the parasitic oscillator, preventing it from functioning. (An experimental alternative is to try several turns of nichrome heater wire without a resistor.)

Usually, a bit of experimentation is required to stabilize radio frequency amplifiers so that reliable neutralization is attained, together with freedom from parasitic oscillations. Much depends upon the physical layout of the amplifier. Other things being equal, one must strive for compactness and short leads. However, proximity between output and input circuits can promote positive feedback, making neutralization difficult. Shielding techniques have to be done carefully so that conductive planes don't actually serve to couple together the very circuit sections they are intended to isolate.

The screen grid electrode of tetrode tubes has as its primary function the greatly reduced internal capacitance between the plate and the control grid. Although the screen grid is polarized with positive d.c. voltage so that it can assist the plate in attracting electrons from the filament or cathode, it is always bypassed by a capacitor to ground radio frequency potential. Thus, it shields or interrupts the electric lines of force from bridging the gap between plate and control grid. In order to allow plate-bound electrons to pass, the screen grid cannot be made from a solid sheet of metal. Accordingly, it is not a perfect electrostatic shield and tetrode tubes *can* self-oscillate under some conditions. This is particularly true because of their high voltage-gain. So, it may be found, depending upon the tube and the layout, that some tetrode amplifiers also have to be neutralized.

Grounded grid amplifiers should, ideally, be free from self-oscillation at the amplifier's resonant frequency. Here, the control grid behaves as an electrostatic shield in much the same way as the screen-grid does in tetrode tubes. Again, however, the shielding action is not perfect and it is not uncommon to encounter self-oscillation in grounded grid amplifiers using triodes. Various remedies will pertain to individual layouts. One may ingeniously have to devise a neutralizing circuit analogous to those used in grounded cathode amplifier circuits. Some other form of negative feedback may do the trick. A small sampling resistance in the input circuit may eat up a bit of drive power, but may justify itself by inhibiting self-oscillation.

It should not be inferred that screen-grid amplifiers and grounded grid amplifiers are free from the parasitic problem. The same remedy applies as discussed for the grounded cathode (filament) triode amplifier and often with less effort.

Allusion was made to parasitic oscillations below the resonant frequency of the tuned amplifier. Let's see how this can come about. For the sake of simplicity, no radio frequency chokes were shown in the grid circuits of the amplifiers of Fig. 6.14. However, when bias networks or bias sources must be applied, radio frequency chokes are then often used and appropriately bypassed as shown in the plate circuits. When such grid-circuit chokes are resorted to, one can then have a tuned-grid/tuned-plate oscillator with the frequency governed by these chokes and their distributed capacitance. (Sometimes, the bypass capacitors do the resonating.) Largely because the inductance of radio frequency chokes is relatively high, the self-oscillation frequency can be much lower than that due to the internal tank-circuits. A worthwhile precaution is to use different-sized radio frequency chokes in the plate and grid circuits. But this is not the only way in which low-frequency 'parasitics' can occur; another oscillatory mode is often encountered in transistor amplifiers.

The operational situation common to bipolar radio frequency and microwave transistors is shown in Fig. 6.15. Note that these devices are worked on the downward slope of their current-gain vs frequency characteristic. As a consequence, transistor amplifiers are 'anxious' to oscillate at a lower frequency, given half a chance. One mode of such low-frequency oscillation can take place if similar chokes are used in the collector and base-emitter circuits. Generally, the oscillation occurs because the transistor 'sees' a tuned collector-tuned oscillator circuit and can supply lots of current gain to make up for conditions that may not be optimum.

Another mode of low-frequency oscillation is a relaxation type and is particularly vicious in that it can destroy the transistor. Such relaxation oscillations readily occur with a poorly regulated power supply. It is as if a

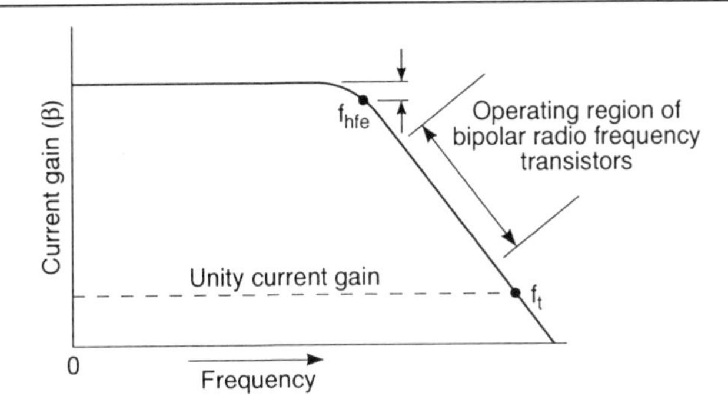

Fig. 6.15 *Current gain vs frequency of bipolar radio frequency transistors. The slope of the gain decline is at least 6 db per octave (20 db per decade). The excess gain at low frequencies makes transistor radio frequency amplifiers vulnerable to low-frequency parasitic oscillations*

physical resistance was connected between the d.c. supply and the transistor amplifier. This simulated resistance forms a time constant with filter or bypass capacitors and our amplifier then operates as a low-frequency multivibrator. When operating in this mode, the tuned circuits play a minor roll; their superimposed ringing can be seen on the sawtooth waveform, but it is the pulsing that dominates the scene. A partial, and often complete, remedy is to provide a tightly voltage-regulated d.c. source, or one having a very low internal impedance. Large electrolytic capacitors connected across the d.c. supply are a step in this direction, but may only serve to alter the time constant, lowering the pulsing frequency. What is usually needed is an RC network, or in some instances, an RLC network.

Such networks take on many forms and are often best determined empirically. (This may be easier said than done because of the danger of losing the transistor in the process.) Generally some series resistance and large shunt-capacitance are called for. There is a conflict between the amount of series resistance and the need for a 'stiff' voltage supply. Nonetheless, one should start with a low-impedance, tightly regulated d.c. supply. The large shunt-connected electrolytic capacitors should be paralleled by smaller ceramic or film capacitors capable of providing good radio frequency bypass action. These ideas are depicted in Fig. 6.16. Hopefully, not all of the precautions will be necessary. Be particularly careful with R_x; if the voltage regulation of the power supply is degraded too much, a time constant will be set up which can provoke the relaxation oscillations. Small values of R_x, however, sometimes prove beneficial. Yet another expedient is to try a grounded base, rather than a grounded (common) emitter amplifier configuration.

Bipolar transistor amplifiers generally do not require neutralizing. However, the application of neutralizing circuitry is easily implemented and

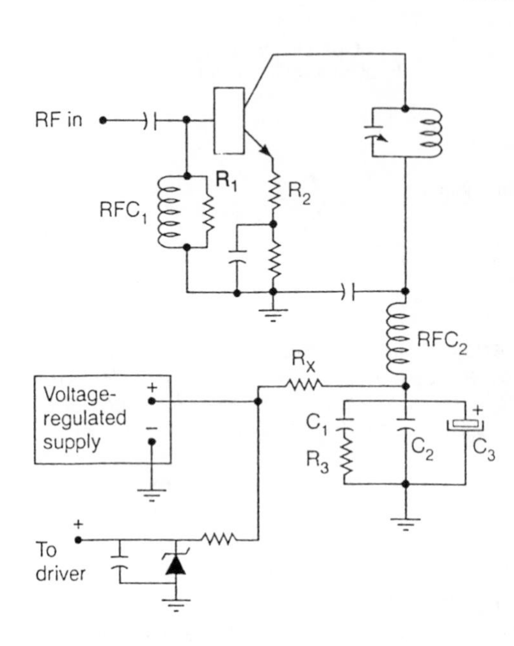

Notes:

Select different inductance values for RFC_1 and RFC_2. R_1 is a *Q* spoiler. Start with 10 **kΩ** and work down. Elimination of RFC_1 may be best approach. Substitute 50 kΩ resistor for start.

R_2 is unbypassed emitter resistance. Several tens to several hundred ohms.

C_1 is often several μF. R_3 tends to be low, several tens of ohms at most. C_2 is conventional RF bypass capacity. 0.01 μF ceramic should be OK. C_3 is an electrolytic capacitor of several-hundred μF.

D_1 is a Zener diode. Together with R_4 and C_4, a decoupling network is formed for the driver stage(s).

R_X in low values sometimes helps inhibit the pulsing tendency. If R_X is too high, the supply regulation will be seriously impaired and the relaxation oscillations may actually be encouraged.

Fig. 6.16 *General guidance to prevent low-frequency oscillations in transistor amplifiers. The relatively high low-frequency gain of grounded-emitter transistors can provoke both LC-resonant oscillations and relaxation-type pulsing*

operates the same way as in tube amplifiers. If such a transistor amplifier exhibits self-oscillation or appears to be close to it, the simpler solution usually is to incorporate some unbypassed emitter resistance. At the expense of slightly more drive, this technique is a clean way to introduce enough wide-band degeneration to stabilize operation. High-frequency parasitics are usually not a problem because of the declining gain vs frequency feature.

Power MOSFET amplifiers can be stable without a neutralizing circuit. Here again, the neutralizing network can be added if self-oscillation is encountered under some conditions. This device, however, is vulnerable to high-frequency parasitics. The most-commonly encountered remedy is to place a ferrite bead over the lead connecting to the gate. This also applies to the JFET. Families of beads most suitable for this purpose not only have a respectable permeability at high frequencies, but are deliberately made lossy in order to absorb energy. A popular ferrite bead is the Ferroxcube 56-590-65/3BB type. Instead of beads, good suppression can often be realized with a non-inductive resistor, commonly in the 10–100 Ω range.

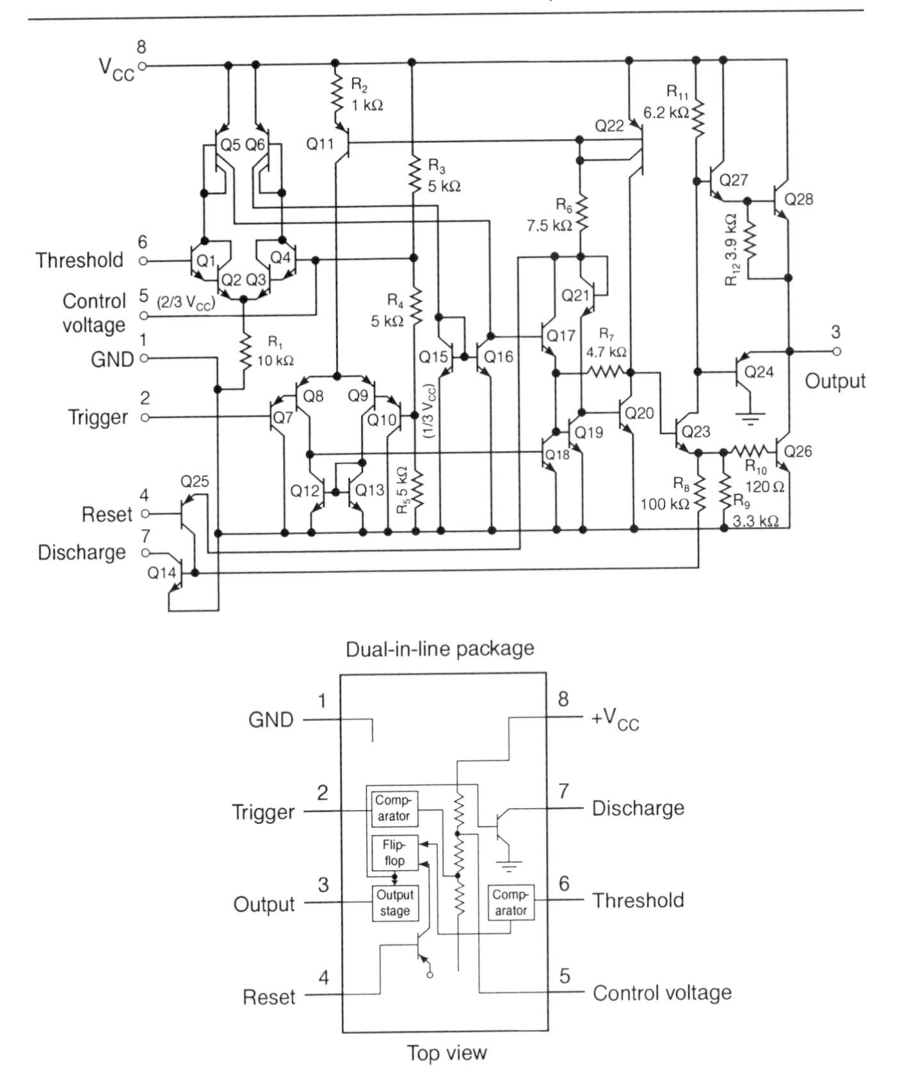

Fig. 6.17 *Circuit and simplified block-diagram of the 555 IC timer. Despite the high number of active-devices, the IC is easy to implement in useful oscillator-related applications*

A curious phenomenon, readily confirmed by electronics practitioners, is that although difficulty is frequently encountered trying to get an oscillator to work, oscillation tends to be obligingly forthcoming from that circuit we've designed as an amplifier. Any reasonable explanation totally eludes the author.

Fancy oscillator functions from the 555 IC timer

The 555 IC is a timing circuit of great versatility in a large number of applications based on control of RC charge and discharge intervals. Because of the association of external RC time-constants and internal active devices, various versions of useful relaxation-type oscillators can be devised. The internal circuitry, together with a simplified block-diagram, are shown in Fig. 6.17. Like many other ICs, the actual circuit tends to be intimidating when one thinks in terms of discrete devices. However, what we have is a simple device with numerous operational modes. This IC has been around for a long time, yet new applications continue to appear in the technical literature.

A great many applications of the 555 timer centre about the basic astable, or multivibrator circuit shown in Fig. 6.18. The reason for this is that by injecting various voltage waveforms into terminal 5, unique output formats are obtained from terminal 3, the conventional output terminal of the device. Although intended for timing applications via charge time of the capacitor C, it is not surprising that numerous oscillator techniques have increased the usefulness of this IC. After all, control of charge discharge times easily leads to repetition of such cycles, or in other words, to oscillation. In essence, the multivibrator of Fig. 6.18 operates much like the classic multivibrators using just a pair of active devices. The salient feature of the 555 IC multivibrator is the readiness in which the multivibrator intervals can be precisely controlled by an injected signal at terminal 5. For the 'plain' multivibrator, terminal 5 is usually bypassed to ground via a 0.01 μF capacitor. The self-oscillation frequency is conveniently plotted against the RC parameters in the graph of Fig. 6.19.

Although, a 50% duty-cycle square wave can be attained via certain modifications, this is not the oscillatory mode of the multivibrator circuit of Fig. 6.18. Indeed, inspection of the arithmetical implication of the formula for duty cycle shows that the 50% condition cannot be attained. Sometimes in the technical literature, a different version of the duty cycle relationship is encountered. Specifically, one sees the formula,

$$D = \frac{R_A + R_B}{R_A + 2R_B}$$

However, this applies when the output at terminal 3 is monitored with respect to the positive side of the d.c. supply instead of with respect to ground. Here again, the arithmetic defeats us when we attempt to attain a 50% duty cycle. With respect to the positive side of the d.c. supply, the duty cycle range is from slightly above 50% to nearly 100%. Conversely, with respect to ground, the attainable duty cycle ranges from slightly below 50% to nearly zero.

By means of a potentiometer and steering diodes, it is possible to keep the

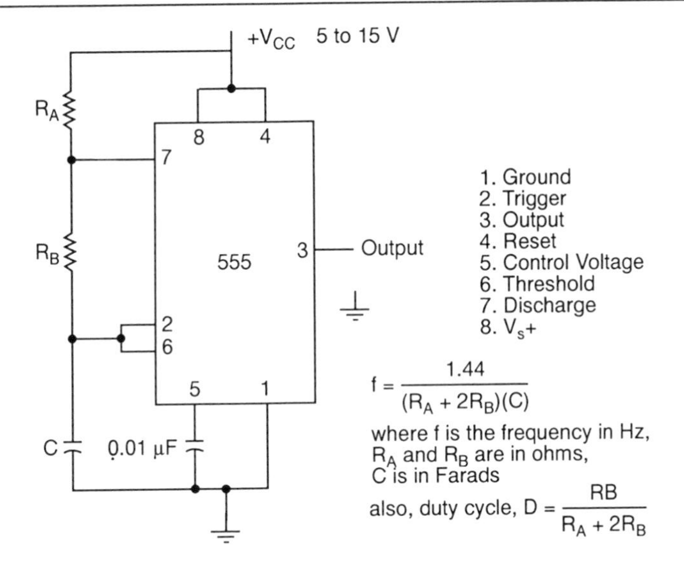

Fig. 6.18 *Circuit of the astable or multivibrator mode for the 555 IC timer. Minor modifications of this basic oscillator leads to various useful applications. By impressing different modulating voltages at terminal 5, the output waveform and frequency can be controlled to produce such functions as frequency divider, pulse-width modulator, pulse-position modulator and emulation of sounds, such as that of a siren*

self-oscillation frequency constant while adjusting the duty cycle. Such an arrangement is shown in Fig. 6.20. Because the total resistance of the potentiometer is fixed at 100Ω, the *sum* of charge and discharge times of capacitor C remains constant regardless of the position of the slider. At the same time, the position of the slider simultaneously changes charge and discharge time, thereby allowing adjustment of the duty cycle while the period, or frequency, remains unchanged. As depicted, different fixed frequencies can be had by selecting different values of capacitor C. The range of adjustable duty cycle is close to 1–99%. This is a valuable feature for a pulse generator, inasmuch as commercial test instruments often lack wide-range adjustment of duty cycle.

The electronic siren of Fig. 6.21 makes use of the alluded control-function of terminal 5. Both IC_1 and IC_2 are self-oscillating multivibrators, but the 500 Hz tone of IC_2 is frequency modulated by the several-second sawtooth wave from IC_1. As can be seen, the conventional terminal 3 output of IC_1 is not used. Interposed between IC_1 and IC_2 is the 2N3702 transistor connected as an emitter follower. This allows minimal loading of the 100 μF timing-capacitor associated with IC_1. The experimenter can readily make use of different audio circuits than the one shown. For example, more

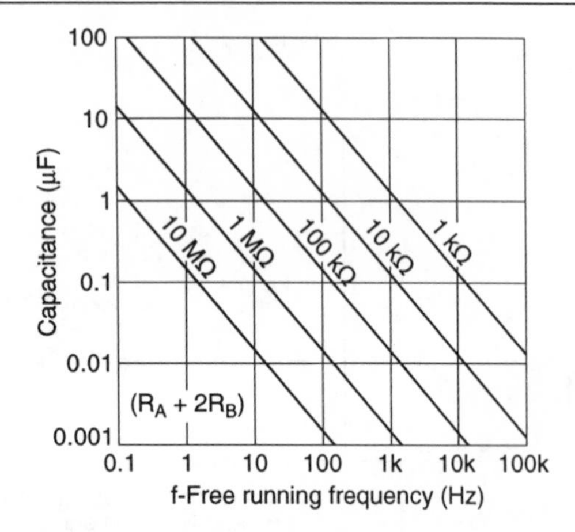

Fig. 6.19 *RC combinations for attaining various frequency-ranges in the 555 multivibrator. It is a good idea to select RC combinations which avoid the use of either extremely high or extremely low values of the resistance or capacitance. Also, if an electrolytic capacitor is used, it should be a high-quality, low-leakage type. Solid-state tantalum types are especially suitable*

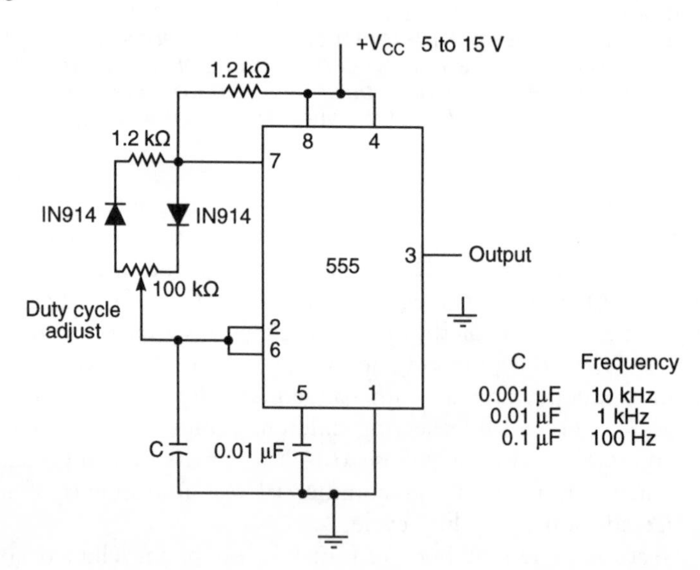

Fig. 6.20 *The 555 timer IC connected to provide adjustable duty cycle at fixed frequency. The arrangement is such that the sum of charge and discharge times of capacitor C remains constant as each is varied by the potentiometer. The adjustable range of the duty cycle is approximately 1–99%*

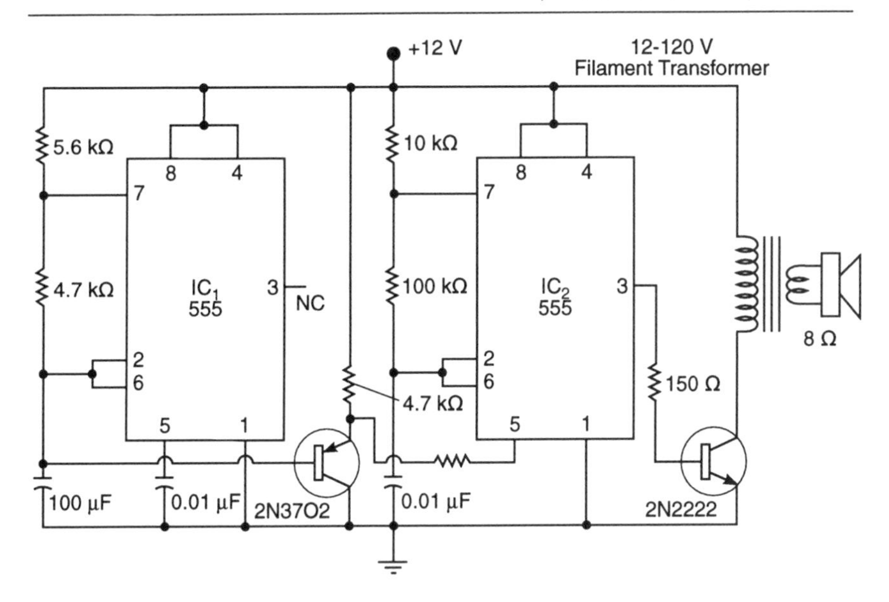

Fig. 6.21 *Dual oscillators arranged to operate as an electronic siren. The low-frequency ramp from IC_1 modulates the higher frequency (500 Hz) oscillator, IC_2. The oscillation frequency of IC_2 is caused to vary above and below its natural 500 Hz frequency, thus emulating the familiar wail of the police siren*

audio power would be forthcoming from substituting a power transistor in place of the 2N2222. An output transformer will not always be required, but is often useful in optimizing power output to the speaker.

Wide tuning range via the difference oscillator

A neat way to design a wide-range test oscillator or signal generator is to utilize the difference frequency between a fixed and variable oscillator. The block diagram of Fig. 6.22 depicts the general idea. The success of this arrangement depends very much on the stability of both oscillators because even slight unintended frequency changes in either drastically affects the difference frequency. That just happens to be the way the arithmetic of the process works. The fixed oscillator is accordingly, crystal stabilized. And, the variable oscillator can be implemented with the various stability techniques commonly used in VFOs. In our favour, is the fact that the degree of stability required in communications work is usually not needed for common test purposes.

Almost any nonlinear device suffices for the mixer and diodes are often used. In many cases balanced or double balanced mixers are desirable

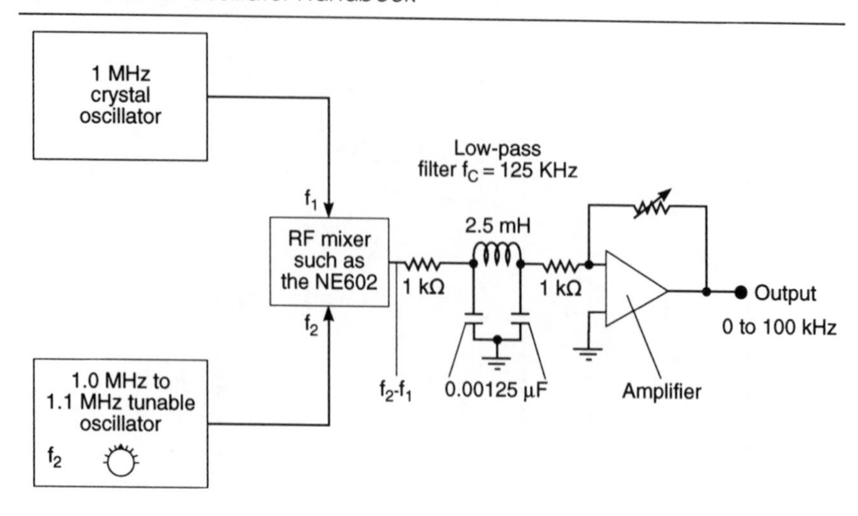

Fig. 6.22 *Basic set-up for a wide-range signal generator. Two oscillators are used. One is fixed frequency, the other is variable frequency. The generated signal is the difference frequency produced by the nonlinear mixing of the two oscillator frequencies. In principle, the output range can extend from zero frequency (d.c.). In practice, 100 Hz or so is readily obtained. Lower frequencies require special attention to the mixer and to the amplifier*

because they relax the requirements of the low-pass filter. If, however, the frequencies of the fixed and the variable oscillator are far removed from the desired output frequency range, a simple single-diode mixer can perform satisfactorily. A suggested IC mixer is the NE602.

The function of the low-pass filter is to remove the oscillator frequencies from the output. Sometimes RC filters suffice, but an LC low pass filter offers better possibilities. To design a single π-section low-pass filter, select a cut-off frequency, f_c, about 10 to 25% higher than the highest planned-for difference frequency. This ensures that the low-pass filter will always operate within the flat portion of its pass-band and will not introduce needless attenuation. Then calculate the L and C elements as follows:

$$L = \frac{R_o}{(\pi)(f_c)}$$ where L is in henries, f_c is in hertz

$$C = \frac{1}{(2\pi)(f_c)(R_o)}$$ where C is in farads, f_c is in hertz

In the above relationships, R_o is the characteristic impedance of the filter and is expressed in ohms. For a wide variety of practical applications, R_o will range from 50 to 1000 ohms. Such LC (image parameter) filters operate best when both their input and output 'sees' the designed characteristic imped-

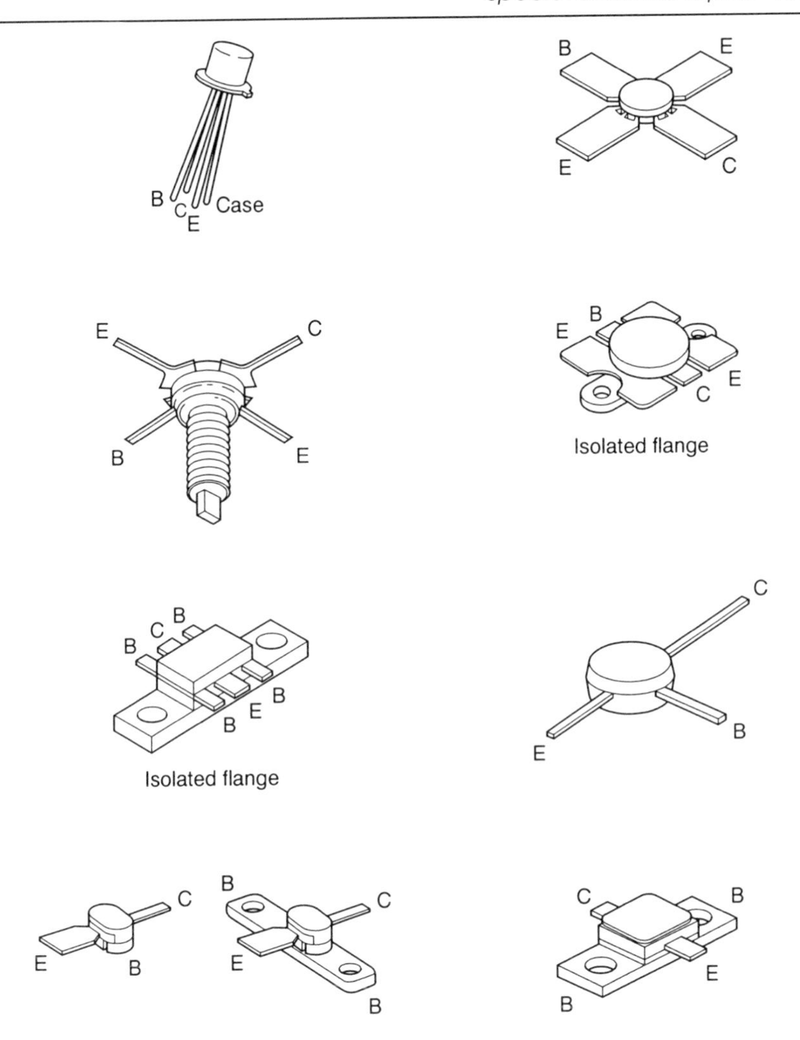

Fig. 6.23 *Unique package styles for UHF and microwave transistors. The primary purposes are to ensure stability in amplifier circuits. However, the packaging has considerable influence on the efficiency of oscillators. Multiple E and B connections reduces lead resistance and inductance. Specific geometric arrangements favour optimum performance in common emitter, common base or common collector circuits*

ance (or resistance). Practical considerations of inductor and capacitor sizes sometimes makes compromises mandatory. Here, a bit of experimentation may be called for. Often, the filter can be made 'happy' by preceding and following resistive attenuators. If these are employed, the attenuation introduced will probably have to be made up in subsequent amplification.

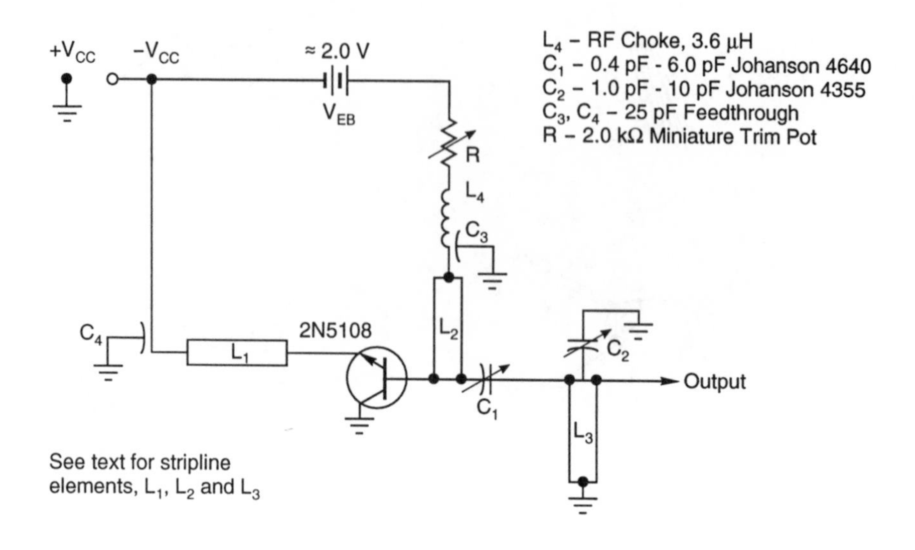

Fig. 6.24 *A 1 W, 1 GHz microwave oscillator. Because geometric shapes and dimensions are involved, it often requires a practised eye to discern the equivalent electrical circuit of microwave oscillators*

Microwave oscillators

Microwave oscillators represent an art and a science unique in electronic technology. This, indeed, is a voluminous subject which justifies treatment separate from the topics selected for this book. However, it will be relevant to at least take a peek at the salient features of microwave oscillators; after all, they still consist of an active device, a resonant 'tank', and a d.c. power supply. (Sometimes, at least two of these basic constituents are contained within one device.)

As we progress to higher frequencies, LC components tend to become vanishingly tiny in physical dimensions as UHF merges with the microwave region of the spectrum. And, ordinary circuit-connecting conductors, together with their environmental items, begin to behave as inductances and capacitances of not-insignificant effect. It turns out that it is better to make use of these reactance effects deliberately, than to ignore, or even compensate for them in conventional LC resonant tanks. Also, best results will inevitably obtain when a dedicated active device is used which the manufacturer has tailored for use in the kind of oscillator circuit we wish to implement. Just using a transistor with a sufficiently high f_T, for example, may not be sufficient for good results. We have to at least also specify the

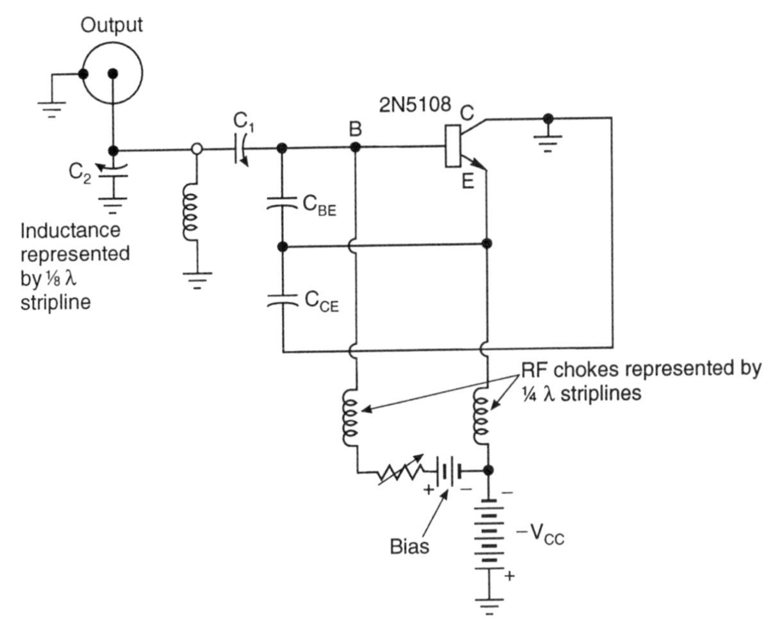

Fig. 6.25 *Equivalent circuit of the stripline microwave oscillator of Fig. 6.24. This is a Clap, series-tuned oscillator of the Colpitts family. Parallel resonance mode operation is also possible, but impedance considerations favour the series mode. Adjustments of C_1 and C_2 are made not only to generate the desired frequency, but to also approach an output impedance of 50Ω*

package type together with the basic circuit it will be used in. Otherwise, lead inductance in the wrong place can easily alter the circuit from what we believe it to be. Fig. 6.23 shows some packaging styles.

Another practical aspect of microwave oscillator design is that we must be prepared for lower attainable power levels as the frequency goes up, if for no other reason, this stems from the ever-smaller physical mass and dimensions of higher-frequency microwave devices. That is why, thermal considerations sometimes override those of a purely electrical nature. For example, transistor heat removal is facilitated by common-collector oscillator circuits; dispensing with the insulating spacer results in more effective heat transfer from the transistor to the metal ground-plane.

A 1 GHz oscillator circuit is shown in Fig. 6.24. From initial inspection, it may not be clear from which circuit family this oscillator derives. It wouldn't be unreasonable to postulate this as evolved from a TGTP circuit of erstwhile tube technology. After all, we see quarter-wave transmission line elements in the input and output of the arrangement. Since such elements can perform as parallel resonant tanks, it appears as we may be on the right track in our analysis.

However, it turns out that this is *not* the operational mode of this oscillator. Rather, we should focus attention on the one-eighth wavelength element. Such a stripline section appears as an *inductance* when its far-end is shorted. This being the case, our circuit becomes the electrical equivalent of a series-tuned Colpitts, or Clapp oscillator. (With the configuration shown, it can also be made to work as a parallel-tuned Colpitts, but this operational mode tends to be discouraged because of impedance levels.) The approximately-equivalent circuit is shown in Fig. 6.25. Note that the capacitive divider is formed by the transistor internal capacities, rather than by actual physical capacitors. In general, microwave oscillators are a study not only of electrical parameters, but of geometric shapes and dimensions.

In order to get the feel of microwave-oscillator technology, it would be instructive to construct the circuit of Fig. 6.24. Use $\frac{1}{16}$-inch Teflon fibreglass stripline-board. Make the widths of elements L_1, L_2 and L_3 $\frac{1}{16}$-inches. (The width governs the characteristic impedance of the lines.) For 1 GHz, the lengths of 'RF chokes' L_1 and L_2 are 2.2 inches, corresponding to $\frac{1}{4}$ wavelength. The length of the 'tank', L_3, is 1.1 inches, or $\frac{1}{8}$ wavelength. This board material has a dielectric constant of 2.55 and this corresponds to a velocity factor of 0.626. Thus, any calculation of line length in air must be multiplied by 0.626. Such calculations will be found to be close, but not exact due to fringing effects, and various proximity disturbances and strays. One thing is clear, however. As we advance in frequency, we must ultimately run out of space as stripline elements become vanishingly small. The region between about 3 and 10 GHz is an awkward one from the standpoint of geometric constraints. Ultimately, microwave cavities prove to be more appropriate resonant circuits, but they too shrink into near nothingness as one gets into the millimetre region.

The 2N5108 microwave transistor is rated at 1 W output at 1 GHz, a respectable power level for this frequency. Yet, there are more capable devices available for the experimenter, and previous records of power vs frequency fall every month. The MRF2016M transistor is rated for 16 W output at 2 HGz. This device is intended for grounded-base circuits, and it would be best to use it in a grounded-base version of conventional oscillator circuits. It has long been desirable to manufacture a solid-state microwave oven operating at the 2450 MHz frequency allotted to such consumer products. At least several hundred watts would have to be developed for a small oven. This has not yet proved to be both technologically and economically feasible.

The Gunn diode

The Gunn diode is a unique microwave oscillating device. In the first place, it is a diode only in the sense that there are two connections to it. Otherwise, it exhibits no rectifying properties. Nor does it contain regions or elements analogous to those of thermionic or solid-state diodes. Superficial inspection indicates a three-section 'sandwich', but the other two of these are there only to facilitate ohmic contact with the inner section where the 'action' takes place. One must not look for any resemblance with pn junction devices, or with Schottky diodes, for there is none. The microwave oscillations that occur do so within bulk semiconductor material, which is gallium arsenide.

For steady-state oscillation, the device is very useful, but also very inefficient. Four or five times more d.c. power must be dissipated that is available at microwave output. For this reason, the device is generally fabricated as part of a relatively massive heat sink. For such continuous-wave operation, the generated frequency is much dependent upon the thickness of the gallium arsenide mid-section. The thinner the section, the higher the 'natural' frequency is. This may be suggestive of piezoelectric phenomena, but an entirely different manifestation of physics is actually involved. Fortunately, by associating the device with a tank circuit or its equivalent, the natural frequency can be 'pulled' considerably. This active inner section of the 'diode' is very thin. When providing such sustained oscillation, the device is said to be operating in its Gunn or transit-time mode.

It should be mentioned that the Gunn diode can also function in another and more efficient mode when pulsed with low duty-cycle bias voltage. Instead of the several-hundred milliwatts available from steady-state operation, peak powers of several hundreds of watts then become available. In this mode, oscillations take place because the device presents a negative resistance to its resonant tank. Relatively efficient oscillation can be obtained far into the millimetre portion of the spectrum. This is known as the delayed transit-time mode and is brought about with higher bias voltage than is used in the aforementioned Gunn mode. The duty cycle will generally be no greater than 0.01%. A good voltage-regulated power supply should be used in either of these operational modes. The bias voltage is a critical parameter for reliable operation.

A basic set-up for utilizing the Gunn diode is shown in Fig. 6.26. This arrangement is particularly suitable for Gunn mode operation, where the indicated dimensions are very important. For such Gunn-mode operation, the manufacturer's intended oscillation frequency and bias voltage must be known before designing its waveguide cavity.

It is important to observe the d.c. polarity requirement of the Gunn diode. This may appear strange in light of the emphasis that the device is not a diode in the usual sense. The paradox is partially resolved by conceding that microwave oscillation will probably take place regardless of the d.c. bias polarity. However, if the 'wrong' polarity is applied, quick destruction from inadequate heat removal is likely. This has to do with the asymmetrical arrangement of the built-in heat sink and with the physics of operation. This

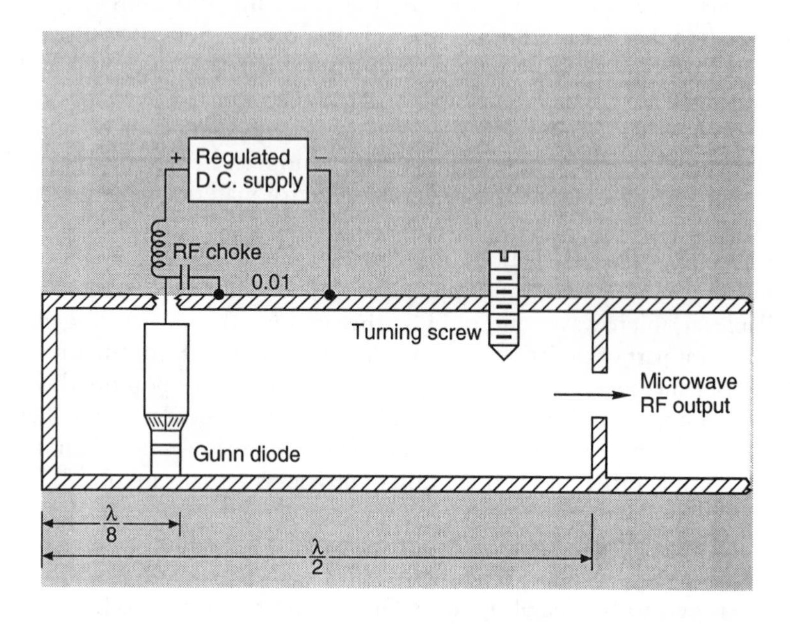

Fig. 6.26 *Basic arrangement for utilizing the Gunn diode. This is similar to the set-up commonly encountered in the popular 'Gunnplexers'. The waveguide section, because of the half-wavelength partition, functions as a resonant cavity. Microwave output power is 'leaked' to the load through an iris in the partition*

is quite complex and was postulated after—not before—microwave oscillations were observed. The internal phenomena in the central 'active' region is somewhat suggestive of the bunching action that underlies the operation of the klystron. As with the klystron tube, most dissipation takes place at the electrode which collects the bunched charge-carriers. In the Gunn device, such collection occurs at the terminal biased with positively-polarized bias voltage.

For operation in the Gunn mode, the device is biased with threshold voltage level. Threshold is simply the bias magnitude at which the device changes from an 'ordinary' Ohms-law resistance to a negative resistance. To make it oscillate in the *delayed* transit-time mode, the pulses of bias voltage must *exceed* the threshold amplitude level.

Note that in the Gunn mode, oscillation could be developed into a resistive load. That is, the oscillations result from action within the bulk material. That is why Gunn diodes intended for higher frequency oscillation in the Gunn mode have *thinner* active regions. Along with the thinner active regions, the higher frequency units require lower threshold bias levels. As previously mentioned, an associated resonant cavity that is tunable can provide a limited degree of frequency adjustment.

By contrast, when performing in the *delayed* transit-time mode, the device behaves more closely to a true negative-resistance and a wide frequency range can be had with a tunable resonant-cavity. (Of course, from a mathematician's viewpoint, *any* oscillator can be represented as a negative resistance associated with a resonant tank-circuit.)

Gated oscillators for clean turn-on and turn-off

It might appear that starting and stopping an oscillator is a trivial matter; in many instances it is, for all we need do is switch the d.c. power supply or somewhat more elegantly, provide for cut-off bias. However, as experimenters and hobbyists ultimately learn, such turn-on, turn-off techniques are not readily implemented without undesirable side-effects. A typical example is the 'chirping' (frequency shift) and key-clicks that occur when the oscillator of a transmitter is keyed. In more serious design work, radar techniques, computer logic and instrumentation circuitry do not operate optimally if oscillator turn-on and turn-off is accompanied by slow rise and fall times, 'glitches', and by interference from noise. All of the alluded malfunctions, and more, are commonly encountered from linear oscillators with LC resonant tanks.

What is needed is an oscillator type with extremely-high gain, very low energy storage, and which operates over the range of cut-off to saturation, rather than in the delicately poised dynamic region of linear oscillators. What we are essentially describing are RC relaxation-type oscillators using logic circuits as active devices. Such topographies are known as gated oscillators and are conveniently implemented with CMOS ICs. The net result is simple, cost-effective and 'sure fire'.

Two such implementations are shown in Fig. 6.27. The circuit of Fig. 6.27a uses NAND gates, while the circuit of Fig. 6.27b uses NOR gates. Operation of these two gated-multivibrators is essentially similar. A practical difference, however, is that the NAND-gate circuit delivers a normally-low output level, and is made active by the application of a sustained logic-1 (high) gate signal to the input. Conversely, the NOR-gate circuit delivers a normally-high output level, and is made active by the application of a sustained logic-O gate signal to the input. Both circuits operate well from a d.c. supply between 5 and 12 V. A voltage-regulated supply is best.

The oscillation frequency for both circuits is about 1 kHz for $R = 680\ k\Omega$, and $C = 1\ \mu F$. Higher oscillation frequencies are then inversely proportional to either of these timing elements. For example, oscillation at about 10 kHz will result by making $C = 0.1\ \mu F$. For best results, R should probably be higher than $60\ k\Omega$, but not higher than 15 to $20\ M\Omega$. The important

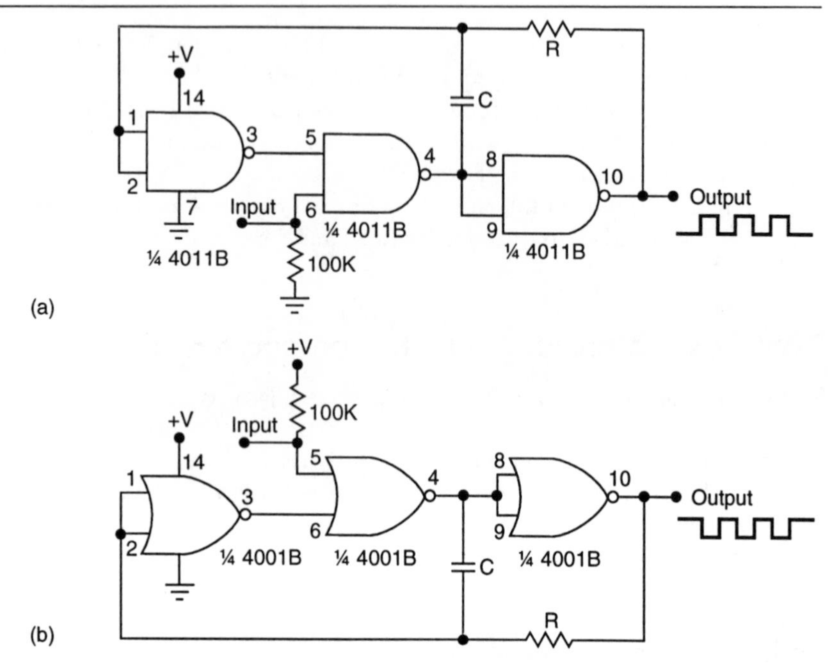

Fig. 6.27 *Examples of gated oscillators. Clean initiation and termination of oscillation is attainable from these astable (multivibrator) logic-circuits. a, Gated square-wave oscillator or clock using NAND gates. The output is normally low. Oscillation commences and is sustained when a logic-1 signal is applied and sustained at the input. b, Gated square-wave oscillator or clock using NOR gates. The output is normally high. Oscillation commences and is sustained when a logic-O signal is applied and sustained at the input*

parameter for timing capacitor, C, is low leakage. Its minimum capacitance will largely be determined by physical aspects of the circuit layout. In any event, cleanly-gated oscillation at 1 MHz by a 100 kHz gating-signal is not difficult to achieve.